Self-Organized Security in Mobile Ad Hoc Networks

Alaa Hassan

United Scholars Publications

United Scholars Publications, USA

www.unitedscholars.net

ISBN-13: 978-0692491867

ISBN-10: 0692491864

Preface

MANETs (***Mobile Ad Hoc Networks***) are ever growing self-organized networks and have applications in industrial, academic and military domains. Routing in MANETs is not trivial because of the constrained resources and infrastructure-less topologies.

AODV (***Ad Hoc On-Demand Distance Vector***) routing protocol is one such protocol which is very popular in MANETs because of its light weight stack and support for low overhead routing. Like other networks, MANETs are also vulnerable to security attacks, which can severely deteriorate the performance and effective throughput of the network.

In this research, we explore AODV routing protocol performance in the presence of two severe security attacks (***Flooding attack and Black Hole attack***). NS-3 (***Network Simulator 3***) is used to simulate a MANET environment with wireless nodes running AODV protocol. Flooding and Black Hole attacks are simulated to show the effects of these attacks on the network in terms of packet drop ratio, transmission delays and routing overheads. Based on these results, we observe that Flooding attack severely depreciates the network performance by dropping valid data packets ranging from 4% to 70% as proportional to the frequency of flooding packets.

Then, we design and build two prevention schemes in order to improve the vulnerabilities observed in the existing protocols. These prevention schemes are inspired by FAP (***Flooding Attack Prevention***) and AMTT (***Avoiding***

Mistaken Transmission Table) techniques which are then simulated in NS-3 and the results are observed. After that, these results are compared in order to analyze the improvements in network performance. Our findings show that the AMTT increases the network throughput when compared with FAP. However, AMTT has a more complex stack and consumes a large amount of processing time and memory. In addition, the transmission delays are severely increased in the case of AMTT. On this basis, we conclude that both of these prevention techniques have their pros and cons and the choice of which optimal technique is used depends on the application scenarios.

The proposed future work is to combine the existing protocols with some medium access schemes to improve the security vulnerability of MANET routing protocols.

Alaa Hassan

July 2015

Table of Contents

List of Figures

List of Tables

Chapter 1: Introduction

In recent years, substantial interest has developed in wireless communication systems. These wireless systems have become popular because of their ability to reach virtually every location on the earth's surface. The global community can now use these networks to communicate and exchange information by using devices such as laptops and smart phones. Therefore, it is hardly surprising that wireless technologies have become a global phenomenon in the realm of personal and business computing [1][2][3][4][5].

MANETs (***Mobile Ad hoc Networks***) are one of the wireless network types [6][7][8]. These networks are infrastructure-less and autonomous networks where a set of mobile nodes are connected by wireless links (see Figure 1) [6][9][10][11][12]. Since the mobile nodes in these networks are autonomous and move at any time and at different speeds, their paths and directions are dynamic and unpredictable (see Figure 2) [6][10][13][14][15][16][17]. MANETs also suffer from the inherit problems faced by mobile and wireless communications, such as power, scarce bandwidth, and transmission quality [18][19][20]. These wireless networks utilize multi-hop relaying and decentralized routing to discover message delivery routes [21][22][23][24][25]. These networks are called challenged networks because of their distributed control dynamic topology [26][27][28][29][30].

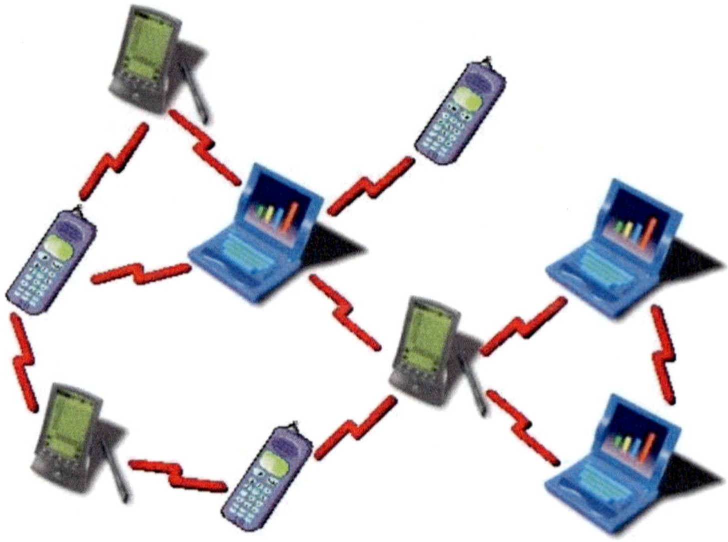

Figure 1: Set of Mobile Nodes [31]

Figure 2: Illustration of MANET Dynamic Topology [32]

Given the widespread use of MANETs, it is necessary to provide a secure method for transmission and communication in these networks [33][34][35][36][37]. The nature of MANETs, as discussed previously, means they are vulnerable to different types of security attacks, such as Flooding attack, Sybil attack, Black Hole attack, Wormhole attack, and Routing table overflow attack [36][38][39][40][41][42].

The design of routing protocols is complex because of the constraints of MANETs. Examples of these routing protocols are AODV (***Ad hoc On-Demand Distance Vector***), OLSR (***Optimized Link State Routing***), and DSR (***Dynamic Source Routing***) [43][44][45]. These protocols have some inherent features to provide security but they are still vulnerable, as ensuring security needs a complete stack at the top.

The protocol suggested for this research is AODV since it is designed to be used in multi-hop wireless MANET scenarios [10][25][45][46][47]. This protocol has been selected because it is an efficient routing protocol which depends on the distance vector approach and it has two mechanisms, Route Discovery and Route Maintenance [9][48][49][50] [51]. These techniques will be explained in detail in Chapter 2 (section 2.2.2).

1.1 Problem Statement

Ad hoc networks are vulnerable to diverse types of attack because of the required cooperation among nodes in these networks. One of the well-known security threats to MANETs is Black Hole attack. In this attack the malicious nodes claim a false short route to the destination to absorb the transmitted data before it is transferred to the destination node, and drop this data instead of forwarding it [52][53][54][55][56]. Previous work has been done to secure MANETs against Black Hole attacks under AODV routing protocol and to elaborate Black Hole attack effects by stating how these attacks disrupt the performance of these networks. However, very little attention has been paid to the study of the impact of Flooding attacks on MANETs using AODV protocol. This attack is also important because of its impact on the service i.e. Denial of Service attack. In this type of attack, a compromised node floods the network with a number of fake messages by sending them to non-existent nodes in the network and this flooding causes network congestion [9][35][57][58][59]. In addition, it can be observed that this attack causes an increase in the routing overhead, average packet drop ratio, and bandwidth requirements in the network. As a result, the performance of MANET will degrade significantly [9][57][60][61]. Therefore, it is important to address this attack as well as its impact on MANETs.

1.2 Research Motivation

Security in MANETs is one of the most important aspects of the basic functionality of networks. Security requirements, such as availability of network services, authentication, authorization, non-repudiation, confidentiality and integrity of the data, can be achieved by assuring that security issues have been met [12][35][58][62][63]. As the communication in MANETs is done by shared wireless media, these wireless networks are highly susceptible to many security threats because of the network characteristics, such as dynamic topology, open medium, lack of central monitoring and management, cooperative algorithms and unclear defense mechanisms [35][36][41][64][65][66][67]. These features have changed the status of Mobile Ad Hoc Networks against possible security threats.

1.3 Aims and Objectives

Flooding attack and Black Hole attack are common attacks which threaten and degrade the performance of MANETs. The major objectives of this research are to represent and simulate both of these attacks by using the network simulator NS-3 and to show their impact on MANETs' performance. This performance evaluation is based on three metrics which are packet drop ratio, average packet delay, and routing overhead.

The aims of this book are:

- To analyze Flooding and Black Hole attacks and their consequences for MANETs.
- To analyze the effect Flooding and Black Hole attacks on MANETs in terms of packet drop ratio, average packet delay, and routing overhead metrics.
- To simulate Flooding and Black Hole attacks using AODV routing protocol.
- To compare the results of both Flooding and Black Hole attacks to find out which is more dangerous to MANETs.
- To discover an optimal technique to prevent the attack identified as the most dangerous.

1.4 Success Criteria

The ultimate purpose of any network is to ensure successful transmission between the network devices in a secure way. It is necessary to investigate the presence of attack and its impact on the network, and the mitigating plans to prevent such attack in the future. Thus, to judge the success of this research, the following questions should be addressed:

Q1: What are the consequences of Flooding attack and Black Hole attack on MANETs?
This question will measure the performance impact on MANETs under normal operation, under Flooding attack and under Black Hole attack.

Q2: Which of these two types of attack are more dangerous to MANETs?
The investigation can be achieved by comparing the results when a MANET is under attack to analyze when the MANET is more vulnerable and which type of attack has a greater impact on it. This investigation can be examined according to a set of parameters such as packet drop ratio, average packet delay and routing overhead.

Q3: Having identified the more dangerous attack, which technique (FAP or AMTT) is better at preventing this attack?
These techniques will be analyzed experimentally. On the basis of this research analysis, it will be easy to suggest the optimal technique that is likely to be used to prevent this attack.

To answer these research questions, it is necessary to model some MANET scenarios with respect to the parameters mentioned above. The proposed routing protocol to evaluate the performance of this network is AODV protocol. MANET scenarios will be simulated based on these parameters and their results will be evaluated in order to answer the required questions.

1.5 Book Outline

This book is structured as follows:

Chapter 2: Related Works
The objectives of this chapter:

- Present an introduction to MANETs with its characteristics, applications, and challenges.
- Explain the routing protocols in MANETs and present an overview of AODV routing protocol and how it can operate on MANETs.
- Investigate the security issues in MANETs by discussing security requirements, security attacks, and security criteria.
- Present the collaborative techniques to prevent these attacks.

Chapter 3: Methodology
The objectives of this chapter:

- Introduce different categories of the research methodology.
- Present an overview of the simulation tool used in this research and the reasons behind choosing this tool by showing the different functions and characteristics that relate only to this research.

Chapter 4: Description of the work
The objectives of this chapter:

- Present details of the simulation topology and its environment.
- Present the different simulation scenarios.

Chapter 5: Evaluation and Discussion
The objectives of this chapter:

- Analyze the effects of Flooding and Black Hole attacks on MANETs and evaluate them in terms of the different metrics.
- Present the analytical study of the extensive simulation to the proposed prevention techniques.
- Summarize the work presented in this research.
- Present the additional recommendation of this research to the future work.

Chapter 2: Related Work

This chapter aims to review the critical points and existing works that are relevant to the research questions of this book and this includes substantive findings as well as theoretical and methodological contributions. It will also present some solutions to ensure MANETs security. This chapter will present related works on:

- MANET challenges.
- MANET attacks.
- MANET security criteria.
- Attack prevention techniques.

A MANET is a refined version of a traditional wireless network since the structure of this network is different from those wireless networks already in use. This network is a system of autonomous wireless nodes connected to each other in a heterogeneous peer-to-peer environment (see Figure 3) [68][69][70][71]. Due to their striking features, MANETs have diverse applications in the real world where the network topology changes rapidly. However, many researchers [72][73][74][75][76] have tried to remove the weaknesses of these networks, such as the dynamic topology that changes over time, limited bandwidth, less battery power of the nodes and security. The reason behind eliminating these attributes is the difficulty in deploying them, particularly when there is an increase in the wireless network size and mobility. It is important to mention that the existing wired networks' security solutions cannot be applied to MANETs [77][78][79][80][81]. Therefore, security is an important aspect in the integrated MANETs to avoid the different types of attacks on these networks. This research will focus on two types of attack, Flooding and Black Hole attacks, as both of these attacks are the most common types of attacks on MANETs and on its access point which is connected to the Internet [82]. The first purpose of this chapter is to review the existing approaches of how these two attacks have an impact on the performance of MANETs. The other purpose is to review the techniques that efficiently prevent, or at least reduce, the consequences of the most dangerous attack on MANETs.

Figure 3: Autonomous Nodes in MANETs [83]

2.1 MANETs (Mobile Ad Hoc Networks)

In recent years, MANETs have been widely used in many civilian and commercial applications, for instance, data collection, search-and-rescue missions, virtual classrooms and conferences where many mobile devices, such as laptops and PDAs, share a wireless medium and communicate with each other [84][85][86]. The reason behind this is not only the availability of low cost mobile devices but also the existence of instant wireless networking capabilities, while this is not possible or is costly in wired networks. MANETs are also self-configuring networks that are dynamically formed by a group of mobile nodes (see Figure 4 and 5) which have no pre-defined infrastructure or centralized management. The nodes in these networks cooperate fully and each one is provided with a wireless transmitter and receiver. This allows the relaying of messages across the network to be delivered to the final destination. Therefore, each node in this type of network must act as a host and a router simultaneously [10][87][88][89][90]. As a host, the nodes in MANETs function as a source and destination, while as a router the nodes in these networks represent as intermediate bridge between the source and the destination and this provides store-and-forward services to all the neighboring nodes in these wireless networks. The reasons behind using the ad hoc networks are clear deployment, speed of development and decreased dependency on the infrastructure [83]. As a consequence, MANETs require the support of multi-hop communications (see Figure 6 and 7). For example, the nodes A and D in Figure 8 can communicate by using nodes B and C to relay the data packets between them. The cooperation between these nodes in MANETs has been monitored by recent protocols and this will be discussed in Chapter 2.

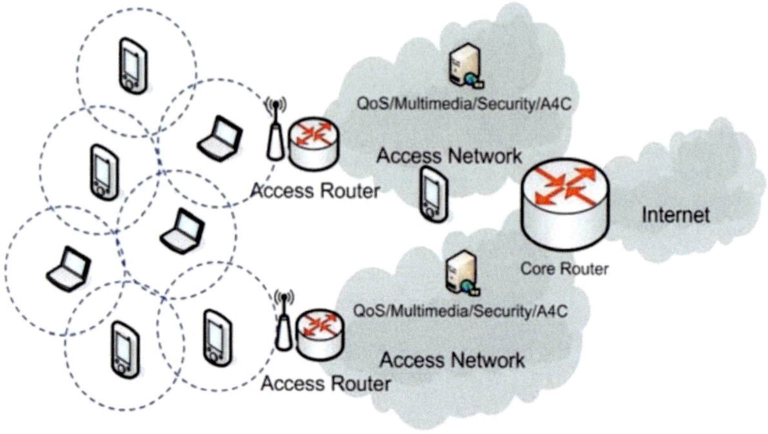

Figure 4: Ad Hoc Network Architecture [91]

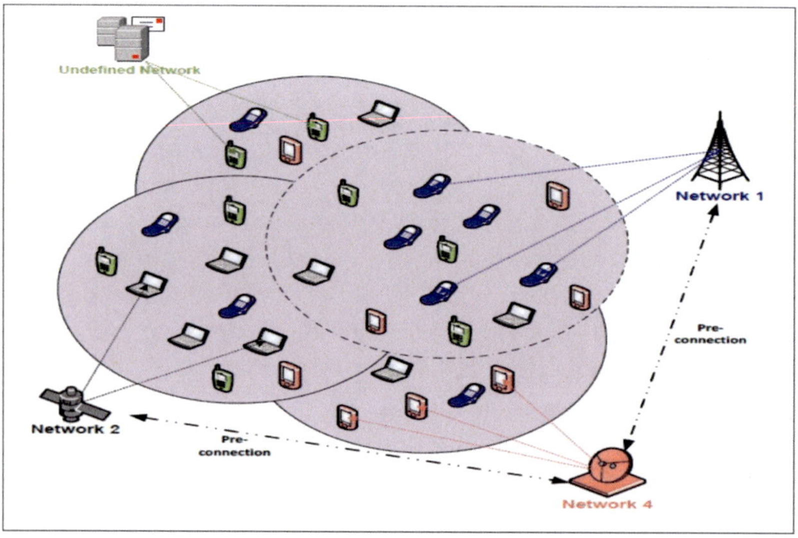

Figure 5: MANETs [62]

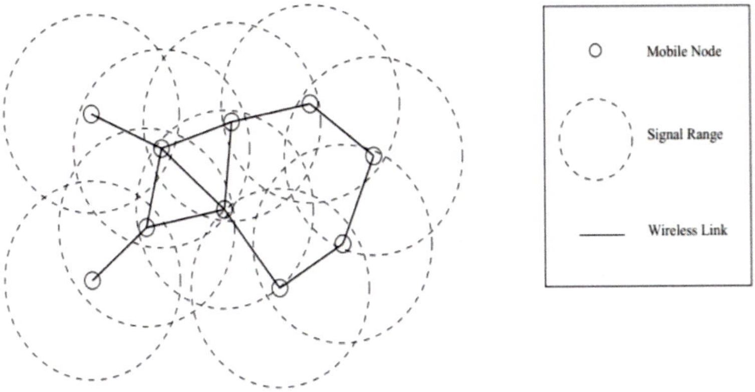

Figure 6: A Multi-hop Communication in MANET 1 [92]

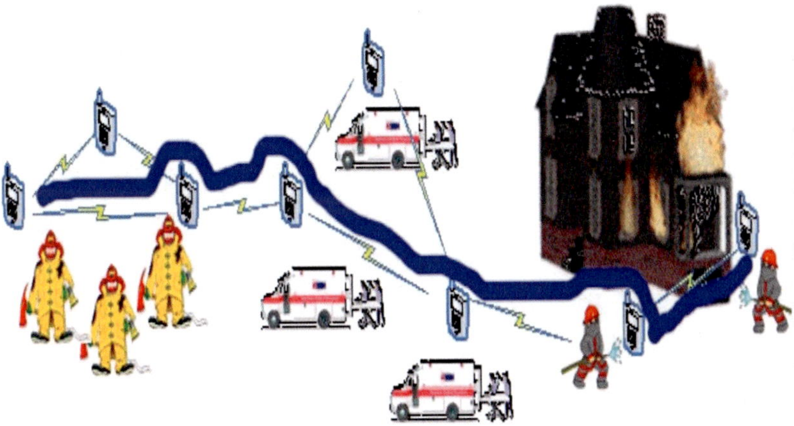

Figure 7: A Multi-hop Communication in MANET 2 [93]

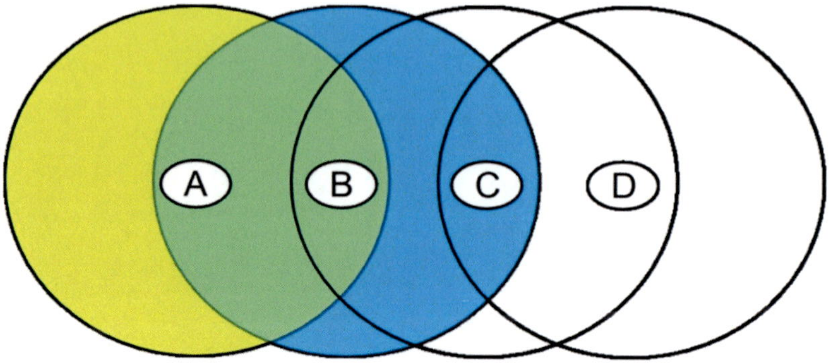

Figure 8: MANET of Four Nodes [94]

The following figure shows some of the major open issues in MANETs:

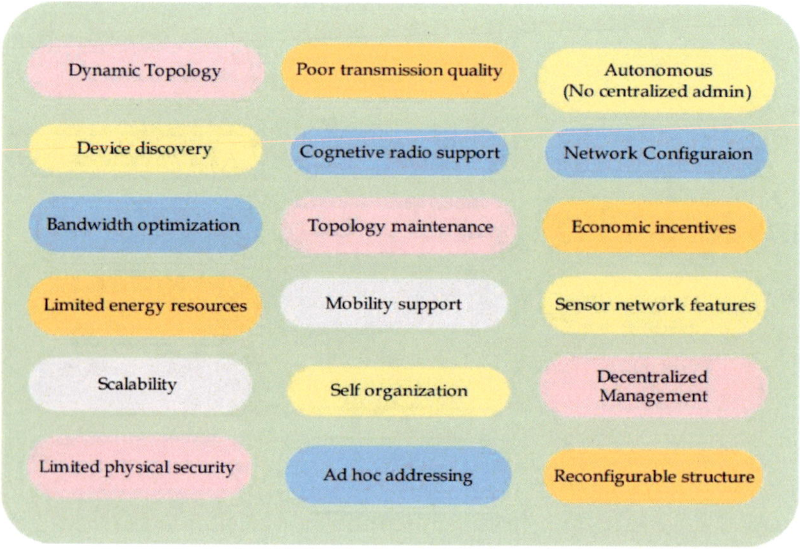

Figure 9: Some Major Open Issues in MANETs [83]

2.1.1 MANET Applications

Figure 10 shows some applications of MANETs and their possible scenarios and services:

Application	Possible scenarios/services
Tactical networks	• Military communication and operations • Automated battlefields
Emergency services	• Search and rescue operations • Disaster recovery • Replacement of fixed infrastructure in case of environmental disasters • Policing and fire fighting • Supporting doctors and nurses in hospitals
Commercial and civilian environments	• E-commerce: electronic payments anytime and anywhere • Business: dynamic database access, mobile offices • Vehicular services: road or accident guidance, transmission of road and weather conditions, taxi cab network, inter-vehicle networks • Sports stadiums, trade fairs, shopping malls • Networks of visitors at airports
Home and enterprise networking	• Home/office wireless networking • Conferences, meeting rooms • Personal area networks (PAN), Personal networks (PN) • Networks at construction sites
Education	• Universities and campus settings • Virtual classrooms • Ad hoc communications during meetings or lectures
Entertainment	• Multi-user games • Wireless P2P networking • Outdoor Internet access • Robotic pets • Theme parks
Sensor networks	• Home applications: smart sensors and actuators embedded in consumer electronics • Body area networks (BAN) • Data tracking of environmental conditions, animal movements, chemical/biological detection
Context aware services	• Follow-on services: call-forwarding, mobile workspace • Information services: location specific services, time dependent services • Infotainment: touristic information
Coverage extension	• Extending cellular network access • Linking up with the Internet, intranets, etc.

Figure 10: MANET Applications [95]

The following figures are some applications of MANETs:

Figure 11: MANET Applications 1 [96]

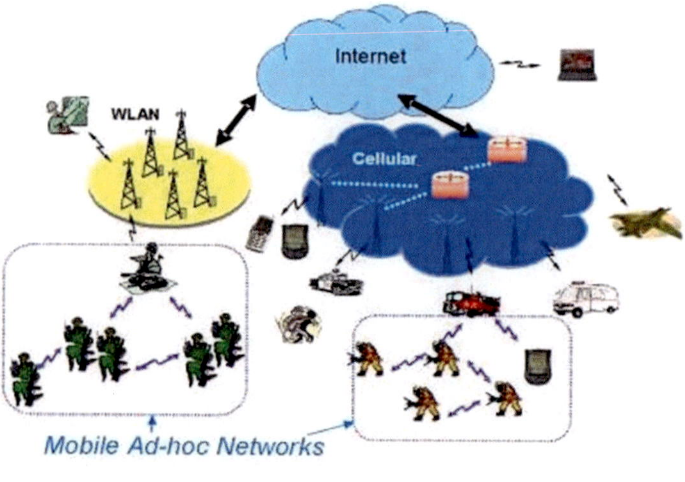

Figure 12: MANET Applications 2 [96]

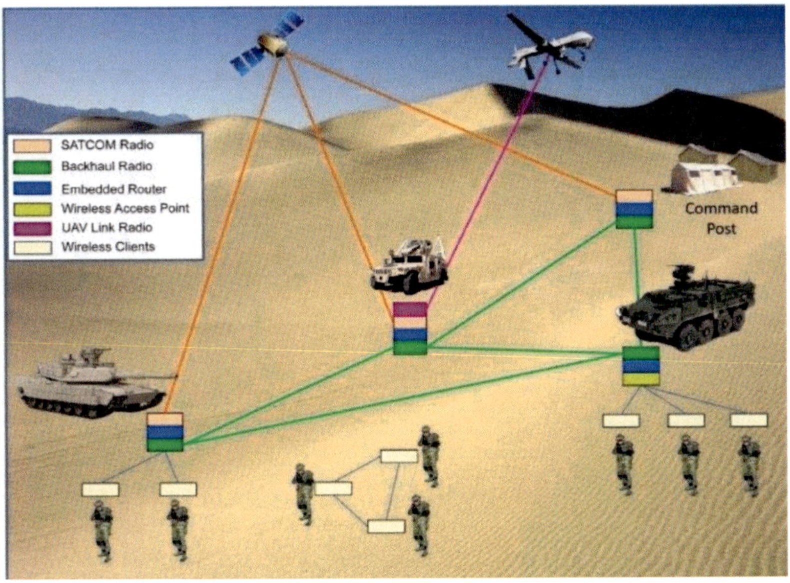

Figure 13: MANET Applications 3 [97]

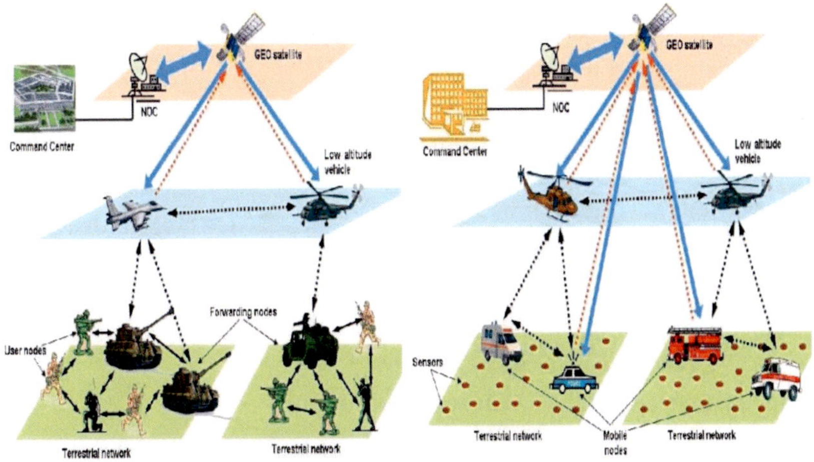

Figure 14: MANET Applications 4 [98]

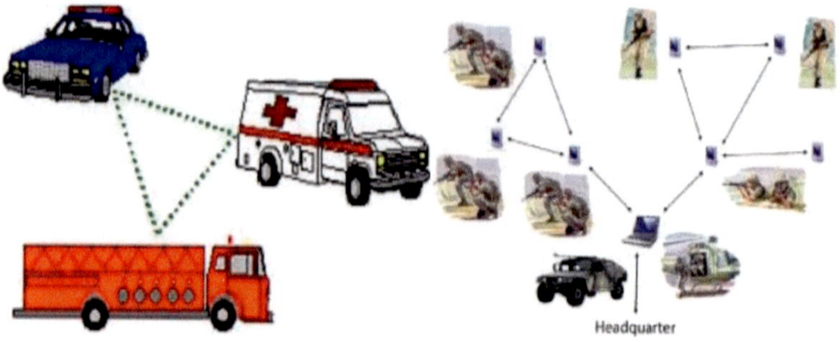

Figure 15: MANET Applications 5 [91]

2.1.2 MANET Characteristics

MANET is important due to its peer-to-peer nature. This means that each peer can self-configure its network card to preserve its connection with the opposite peer. The individual nodes in these networks can behave as a router to forward data by sending and receiving packets to and from other nodes in the network. Therefore, MANETs have become one of the fastest growing areas of research because of the rapid proliferation of cellular phones, notebooks and other wireless devices that have wireless cards installed in them.

In addition, one of the distinctive features of MANETs is being decentralized compared to the usual networks, such as LAN and WAN, and this characteristic makes it suitable since its central nodes cannot be relied on and it does not necessitate a pre-existing communication infrastructure for operation. MANETs are also suitable for emergency situations, such as military conflicts and natural or human-induced disasters, due to minimal configuration, fast deployment, and no pre-set topology being required.

Furthermore, MANETs have several salient features that make them different from other wired and wireless networks, such as:

- **Dynamic Topologies**:

  The nodes in MANETs can move arbitrarily and this means that they can connect to the network or leave it at any time. As a result, the topology of these networks can rapidly and randomly change at unpredictable times as shown in Figure 16, and they may also be composed of both bidirectional and unidirectional links [12][62][99][100][101][102][103].

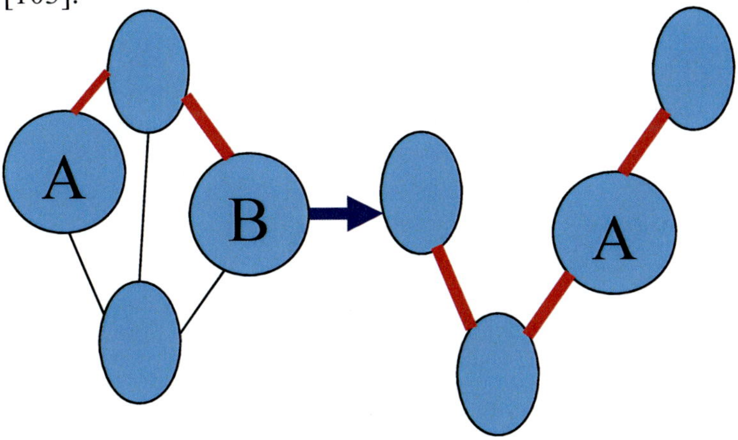

Figure 16: Topology Change

- **Energy-Constrained Operation**:

  In MANETs, some or all of the nodes keep their energy depending on batteries or other exhaustible means. The significant optimisation criterion for these nodes is energy conservation [12][25][99][100][101][102][103].

- **Limited Bandwidth**:

  The reliability, efficiency, and capacity of wireless links in MANETs are often inferior compared to the wired links. This is due to the fluctuating link bandwidth of these wireless links [12][62][99][100][101][102][103].

- **Limited Physical Security**:

  MANETs are more susceptible to security threats than wired networks. Spoofing, jamming, eavesdropping, and DoS attacks are examples of these security threats. Many security techniques have been applied to wireless networks to decrease these threats. It can be noted that the decentralised nature of MANETs has the ability to provide more robustness against single failure points [12][62][99][100][101][102][104].

2.1.3 MANETs Challenges

MANETs have been an interesting area of study over the last few years. However, there are many current research challenges faced by using these networks, and these challenges are:

- Scalability
- Quality of service
- Client-server model shift

- Security
- Energy conservation
- Node cooperation
- Interoperation

2.1.3.1 Scalability

A routing protocol ability to scale is to perform efficiently as the network grows to be large in numbers. MANETs suffer from many scalability problems in capacity and this can be exemplified by simple interference studies. It can be observed in a non-cooperative network that there is a decrease at a rate $1/\sqrt{N}$ for a limited throughput per node where N is the number of nodes. When a network has 100 nodes, a single device can obtain roughly one tenth of the network data rate and this problem cannot be set except by physical layer improvement, such as by use of directional antennas. There are examples that necessitate large overhead and abundant control traffic as the network size grows. Consequently, scalability is an interesting issue and it is important to consider it in the design of MANETs [78][79][95][105][106][83][107].

2.1.3.2 Quality of Service (Qos)

QoS is a set of service requirements which must be met by the network while transporting a data packet from a source to its destination. To ensure QoS in MANETs is very complicated because of the dynamic change in the network topology and the imprecise of its accessible routing information. This problem is faced by MANET designers as they need to provide the best effort service for many applications that have diverse requirements, such as live video, voice and file transfer. Figure 17 depicts some challenges when proving the QoS in MANETs.

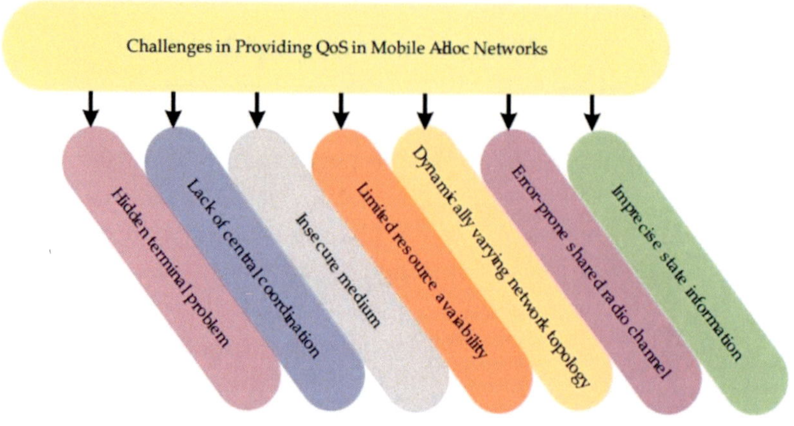

Figure 17: Some Challenges when Proving the QoS in MANETs [83]

QoS solutions are expanded to meet the requirements in wireless environments. These requirements could be defined by a wide set of metrics including jitter, delay, packet loss probability, bandwidth, and so on. Although there are great efforts in the QoS area, it is difficult to achieve the required QoS in MANETs. The reasons behind this problem are due to MANETs' features and constraints, such as limited bandwidth, dynamic topologies, and variations in link and node capabilities [79][83][92][103][107][108][109].

2.1.3.3 Client-Server Model Shift

A network client is set up to use a server and this can be found automatically or by a static configuration. There is no need in the structure of MANETs to collect IP addresses into subnets because there are no servers, but the basic services still exist, such as name resolution, address allocation, and authentication.

In MANETs, a different addressing approach might be required because of the node mobility and the infrastructure-less nature of these wireless networks. Moreover, it is not obvious who is responsible for managing the different network services. Although there are many researches in this area, the problem of shifting from the conventional client-server model still needs to be properly addressed. Fig 18 illustrates the client-server model [6][107][110][111].

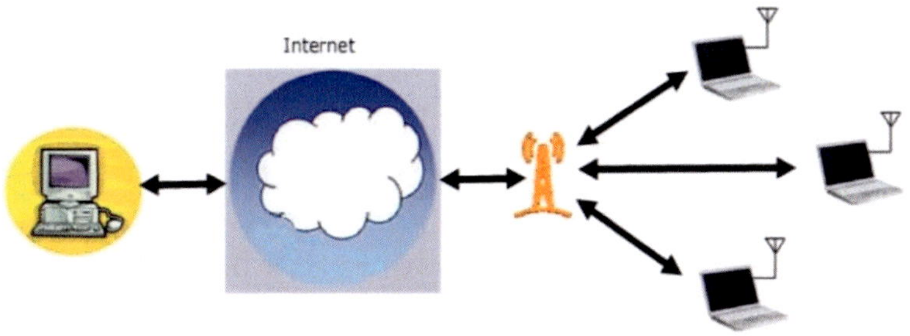

Figure 18: Client-Server Model [107]

2.1.3.4 Security

Security is a vital issue that has to be addressed in relation to MANETs since there are many applications, such as military and confidential meetings, that need a high degree of security against various active and passive attacks. MANETs have no clear line of defense, so, they are particularly prone to network users and malicious attackers. The dynamic network topology in these networks is vulnerable to infiltration, eavesdropping, and interference due to the lack of certification authority or centralized network management. Given this evidence, it can be observed that security is the main block to the commercial applications [12][24][62][78][107][110]. Therefore, it is necessary to indicate some key security requirements in MANETs specified below. All of these will be defined in section 2.3.1:

- Authentication
- Authorization
- Availability
- Non-repudiation
- Confidentiality
- Integrity
- Privacy

2.1.3.5 Energy Conservation

Energy conservative in MANET research is of paramount importance since the nodes in these networks usually have a very limited energy supply. There are two main identical research topics to address this problem, lifetime maximization of a battery and lifetime maximization of the network. The former is related to the node cooperation issues and the commercial applications. The latter is essential, for example, in military environments where it assumes node cooperation. These goals can be achieved by developing better batteries, and also by making a terminal's operation of the MANET more energy efficient. The first approach may increase battery life by 40% (with Li-Polymer batteries). It is necessary to achieve energy savings through a range of techniques, such as flash memory, variable clock speed CPUs, and disk spin down because of the device power consumption. The second approach focuses on the network interface of the device and this is the major consumer of power. In this approach, the Energy efficiency can be enhanced by expanding transmission and reception technologies on the physical layer.

It can be observed that much research has been achieved at the physical, MAC (Medium Access Control), and routing layers and a little has been done at the transport and application layers. Nevertheless, there are many studies on these two layers to be executed [107][110][112][113].

2.1.3.6 Node (MH) Cooperation

As far as the security and routing problems are concerned, the cooperation among nodes stands in the way of commercial applications. There is no alternative to receiving the corresponding services from others, but one has to rely on other people's data. The problem in the node cooperation is the difference in the data amount and priority, so the situation becomes too difficult. The fire alarm box does not waste its batteries to relay the data, and it prevents them from accessing the other nodes due to the restrictive behavior. The cooperation of nodes may lead to the introduction of billing and this problem is similar to the Internet congestion control. Malicious users might be charged high rates and well-behaving network members might be rewarded. On the other hand, any billing mechanism implementation is still challenging. These concerns are still wide open [107][110][114][115].

2.1.3.7 Interoperation

Interoperation is also an important issue in MANETs and it aims to enable these networks with potentially heterogeneous networking technologies to correspond with each other while preserving the administrative management of individual networks. The problem of interoperation in these networks is when two autonomous MANETs move into the same area, thus, the interference with each other becomes unavoidable. On the other hand, the concern of connecting two MANETs is not trivial and these networks might use different synchronization, MAC or routing protocols. As a result, it is important to consider this research topic since it has many implications at all levels of system design [107][110][116].

As far as the mentioned points are concerned, security turns out to be a major concern as shown in Figure 19.

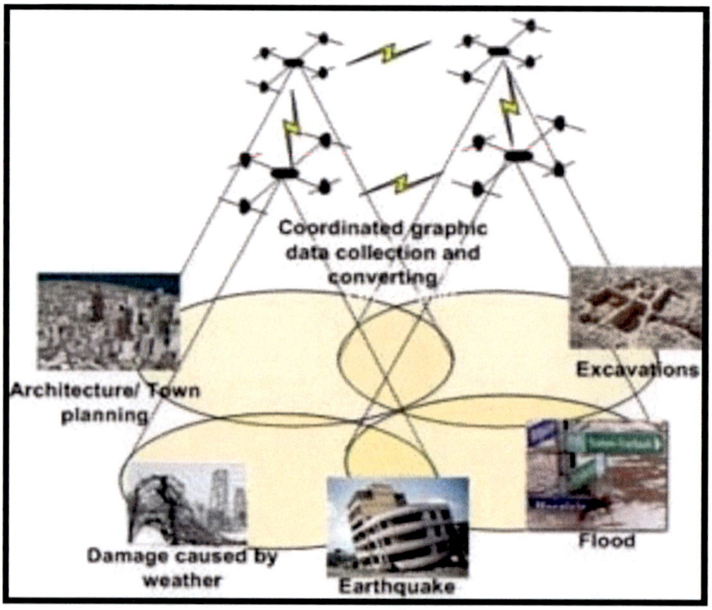

Figure 19: Scheme of the Importance of Security [117]

2.2 Routing in MANET

Researchers have paid a great deal of attention to develop an efficient and secure routing protocols for MANETs. The reason for this attention is because of the routing problems in ad hoc networks. With this problem, the network has difficulty reaching the loop-free, low-overhead, open peer-to-peer network and adapting to the dynamic topology of the wireless network. This is not a trivial task to select an optimal path, when no central node is responsible for keeping and managing the network. To provide a secure routing protocol for these networks, there are many requirements such as a guarantee of correct route discovery, detection of malicious nodes, stability against attacks, and confidentiality of network topology[83]. Figure 20 displays the security issues in each TCP/IP layer [83]. The aim of a MANET routing protocol is to provide paths that are not only optimum in terms of some standards (minimum distance, maximum bandwidth, and shortest delay) but also in satisfying some constraints, for example, the limited power of mobile nodes and the limited capacity of wireless links [99][118][119]. Because of the continuous movement of the nodes in MANETs, the need for an efficient routing protocol has increased and this allows each node to discover its own view of the whole network topology with the least overhead possible as shown in Figure 21.

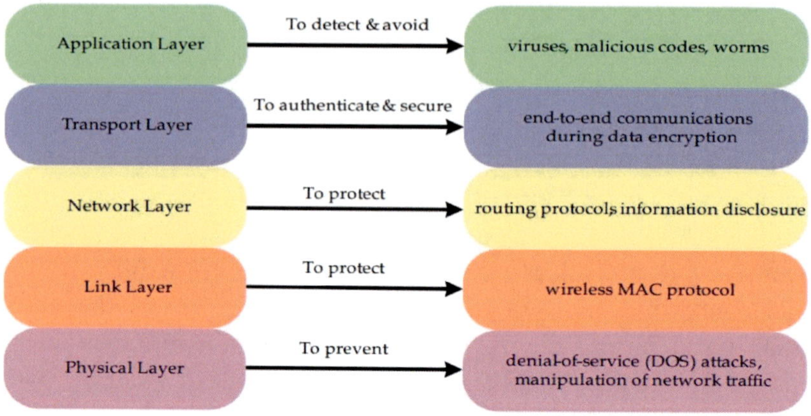

Figure 20: Security Issues in TCP/IP Layers [83]

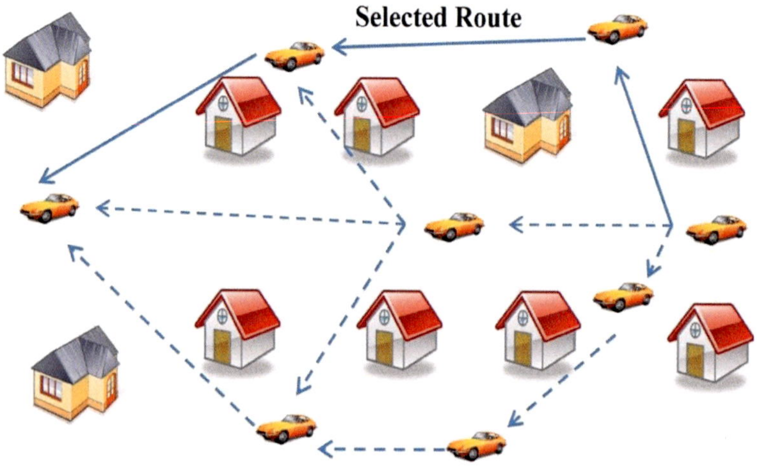

Figure 21: Illustration of the Movement of Nodes by Using an Efficient Routing Protocol [120]

In MANETs, routing protocols can be categorized depending upon their communication model, network structure, state information, and routing strategy as shown in the following figure, but most of these routing protocols depend on their network structure and their routing strategy [121][122].

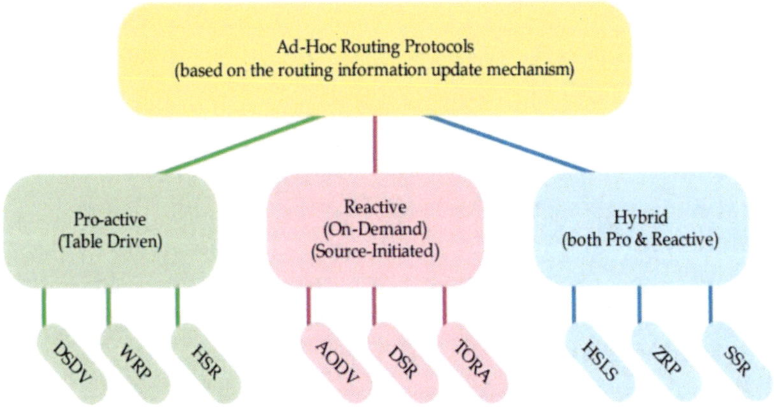

Figure 22: Ad-Hoc Routing Protocols Classification [83]

Routing protocols in MANETs, depending upon their routing strategy, have been classified in to three types [99][123][124][125][126][127]:

- **Proactive Approach (Table Driven Routing Protocols):**

  Proactive routing protocols are able to maintain, periodically, the routing information and store it in the routing tables. In this class, there are many differences among the protocols, such as number of tables, routing structure, and frequency of updates. Consequently, each protocol in this class has different reactions to the changes of topology. The mechanism used to update and discover routes in this class is flooding of routing information. Because of the periodic updating in the proactive protocols, they are able to overhead and control traffic more than the other classes. This increases when there is a growth in the number of nodes. In addition, these protocols have an extensive use of memory to store those tables. WRP (***Wireless Routing Protocol***) and FSR (***Fisheye State Routing***) are examples of proactive protocols. With WRP [99], four routing tables can be maintained by each node. When the network expands, this protocol consumes a significant amount of memory. To guarantee the connectivity of neighbors, 'Hello' messages are used and this consumes more power and bandwidth. The other proactive protocol is FSR [99] which is able to update network information for nodes that are within its scope. As a result, although FSR is not adaptable to high mobility, this protocol is more scalable than WRP.

- **Reactive Approach (On Demand Routing Protocols)**:

Reactive protocols have the ability to maintain or discover a route and this leads to reduce the overhead created by these protocols [99]. The flooding strategy is the strategy used for route discovery. The reactive approach is classified into source routing and hop-by-hop routing. In the first type, the headers of the data packet can transfer the packet to the destination but the intermediate nodes are not able to maintain the routing information. These protocols have a high level of overhead when there is an increase in the number of intermediate nodes. Furthermore, these protocols might be exposed to a route failure. In the second type (hop-by-hop routing), packets of these protocols carry only the next hop and destination addresses. This means that the routing information of active routes can be maintained and stored by the nodes. The disadvantage of these protocols is their delay caused by the route discovery process. Examples of these well-known reactive routing protocols are AODV (***Ad hoc On-Demand Distance Vector***), DSR (***Dynamic Source Routing***), and LAR (***Location-Aided Routing***) [99][128]. AODV protocol is a hop-by-hop routing protocol and can discover and repair routes more than the DSR routing protocol because of its dynamic strategy. This protocol has the ability not only to evade the infinite loops problem using DSRs but also to reduce control traffic and overheads by maintaining only the active routes. Therefore, AODV routing protocol is appropriate for node mobility, loads, and density. DSR is a reactive routing protocol [128][129] and by using the approach of route discovery and maintenance, it can discover routes on demand. So, in order to increase robustness and attain load balancing, multiple routes are applied. DSR has a well operation with high mobility nodes because this protocol can recover from route failure rapidly. The reason behind this protocol's capability to operate well over a medium network density is that it supports up to one hundred nodes. LAR can detect the location of nodes by using GPS information and this reduces overheads because of the flooding. In this protocol, there are two classifications for route discovery. The first classification can limit the propagation of a RREQ for a define area (Request Zone). The second classification can store the destination coordination when it receives the RREQ packets and this reduces overheads; however, there should be a GPS for each node.

- **Hybrid Routing Protocols:**

These protocols are a collection of reactive and proactive protocols. The proactive approach function is to maintain and discover routes to the neighbor nodes, whereas the nodes which are far away from its routes are discovered reactively. Subsequently, proactive and reactive protocols can respectively minimize overheads and delay. These protocols are more scalable than the other nodes participating in routing and topology discovery. For example, the nodes in ZRP (***Zone Routing Protocol***) are gathered into zones, and they can connect with each other depending on their locations in the zone. The other example is SLURP (***Scalable Location Update Routing Protocol***) and this is able to adapt the changes in node mobility and density. To manage node location and eliminate global routing, these protocols use the information of GPS. Each node has a home region but when it moves to a new location, it sends the new location to its home region. Thus, the source node requests the home region of the destination when a route is required. As a result, hybrid protocols are optimal for huge networks where there is a high number and mobility of nodes [128].

2.2.1 AODV (Ad Hoc On-Demand Distance Vector Routing)

This section presents an overview of AODV protocol and how it can be operated in MANET. This protocol allows the nodes in a MANET to be self-starting, dynamic, multi-hop routing [91][130][131]. AODV [122] is a reactive routing protocol because it can determine only on demand a route to the desired destination. AODV [132] allows the nodes of MANETs to recognize very quickly the link breakages and also any changes in the network topology. It involves a wide range of operations, such as, route table management, path discovery, local connectivity management, and path maintenance [133]. All of these operations are not loop-free, but they might converge in a highly dynamic network by avoiding the 'counting to infinity' problem identified by Bellman-Ford, which calculates the shortest path to the destination in the network [91].

When there is a break in any link, the role of this protocol is to notify the other nodes to avoid the routes using the demolished link. One of the key design elements for this protocol is DSN (***Destination Sequence Number***) [134]. Each entry in the routing table has its own sequence number for each participating node in the MANET. DSN has the sending information from the destination to a requesting node and it guarantees the 'loop-freedom' which is easy to implement [132][135]. Since there are many routes to the destination, the requesting node has the capability to select the route with the highest sequence number.

AODV routing protocol is designed to be used in multi-hop wireless MANETs which include tens to thousands of mobile nodes, and its aim is to provide secure and reliable data transmission over these networks. This protocol is proposed to be utilized by these networks since the nodes in these wireless networks can trust each other by using preconfigured keys. In addition, this protocol has been recommended, not only to reduce the broadcast of control traffic, but also to avoid overhead data traffic with the aim of improving the scalability and the performance of ad hoc networks [132][136][137]. AODV protocol has been described according to the RFC 3561 [132]. This section will introduce an overview of the protocol operation as well as an explanation about the three message types of this protocol. The effects of different types of attack on MANETs by using AODV routing protocol can be observed within this research.

2.2.2 Protocol Overview

The AODV protocol has three varied types of messages, such as RREQs (Route Requests), RREPs (Route Replies), and RERRs (Route Errors) [138][139]. The nodes in MANETs have a capability to exchange these messages via the UDP (***User Datagram Protocol***) [132][139]. After that, the IP (***Internet Protocol***) header processing is applied, such as TTL (***Time_To_Live***) and this is used to manage the life time of messages issued by any node [132]. For example, a requesting node must always use its IP address as the originator IP address for the messages. These messages might be propagated by the use of IP limited broadcast address (for example, 255.255.255.255 for IPv4). This indicates that these types of messages are not just blindly forwarded. On this basis, it may be inferred that the operations of AODV necessitate different messages, for example the RREQ message, to be largely used in MANETs. The propagation of RREQs is declared in the header of the

IP message by the TTL; however, the fragmentation of the IP packet is not required [140].

If there are one or more active routes between the source and destination nodes, the AODV protocol will be passive. However, it will be active, when there is a route to a specific destination [141]. The node running AODV broadcasts a RREQ when there is a required route to a new destination. This route is identified either when the destination receives the RREQ messages or when there is a located 'fresh enough' route from the intermediate nodes to the destination. The term 'fresh enough' is a legal route to the destination whose sequence number is greater than or equal to the sequence number contained in the RREQ [52][132][142]. On receiving the RREQ, a RREP message is unicasted back to the RREQ message originator through the received request. Then, the source node will start to send the data packets. When the source receives many RREP messages, the route which has the shortest hop count is selected. It can be observed that the receiving nodes cache a route back to the sender in its routing table. The RREP message can be unicasted from the destination or from any intermediate node to satisfy the request.

The link position of the next hop can be observed by each node in the MANET in an active route. When there is a break in any link, all the nodes will be notified about this through a RERR message [141]. Then, the RERR message lists the unreachable destinations due to the broken link. This can be done using a 'precursor list' maintained by each node. This list preserves the neighbouring node's IP address which forward a RREP [143]. Hence, it can be said from the mentioned procedure that the information of the 'precursor list' can be straightforwardly attained from the generation of a RREP message. The processing of RREPs gives routes to subnets, which means that a RREP contains a nonzero prefix length, when the source of a RREQ message is inferred among the precursors to the subnet route but not to the specific destination [132].

A node using the AODV protocol can also send a RREQ for a multicast IP address. For example, the source of a multicast IP address may have to follow specific rules. It is necessary to correct multicast operations by intermediate nodes which are not permitted to be source or destination nodes for IP multicast addresses. Nor can they provide for any special multicast protocol processing. For such multicast-unaware nodes, the multicast process as a destination IP address must be performed in the same way as for any other destination IP address.

As previously mentioned, the routing table contains information about short-lived routes. For example, it has completed reverse paths to the *originating node*. Figure 23 shows the AODV fields with each routing table entry [141]:

Destination IP Address	DSN	Valid DSN flag	Other flags			Network Interface	Hop Count	Next Hop	List of Precursors	Lifetime
123.456.789.012	12	0	0	1	0	lo0	5	456.789.012.345	...	4
321.654.987.210	45	1	1	0	1	le0	9	456.789.012.345	...	6
456.789.012.345	13	0	1	0	0	lo0	3	123.456.789.012	...	7

Figure 23: Example of an AODV Routing Table [141]

One of the most significant fields in AODV protocol is DSN since the l*oop freedom* is *assured,* particularly if the destination becomes unapproachable when there is a deactivated or broken link [144]. In this case, not only the route will be invalidated, but also the routing table entry state is inoperative.

2.2.3 AODV Operation Example

The importance of this protocol is due to the presence of Route Discovery and Route Maintenance mechanisms [9][48]. Route Discovery depends on the RREQ (***Route Request Message***) and RREP (***Route Reply Message***). [9][141] have proposed that the route information of intermediate nodes is stored in the routing table entries. In this mechanism, the RREQ message is broadcasted by the node to provide a route to the destination. When the destination receives this RREQ, it will send a RREP message to the source of the RREQ. [135] have suggested that this mechanism prevents routing loops by using the sequence numbers in each routing table entry. This can be helpful to determine whether the routing information in these tables is up-to-date. However, Route Maintenance depends on the RRER (***Route Error***) message and this can handle the dynamic network topology in MANETs. [9][141] have proposed that this mechanism uses the RERR message to maintain the routes by sending notification to the other nodes about a link failure. AODV operation does need these types of messages to be disseminated [144]. The following figure shows example of an ad hoc network. This network has 14 active nodes, the source is node 1 and the destination is node 14. Node 1 will broadcast a RREQ message through the network until it is being delivered to the destination.

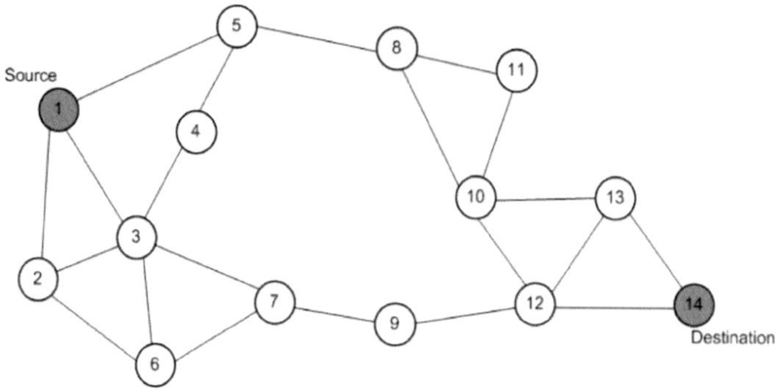

Figure 24: Ad Hoc Network Example [141]

2.2.3.1 Route Request – RREQ

A RREQ message is broadcasted from node 1 as a flood request to its neighbors (node 2, node 3, and node 5) in the network with the request of the destination node 14 and this is illustrated in Figure 25. In this step, the sequence number of source node 1 is incremented and it creates a DSN, originator IP address, destination IP address and OSN (originator sequence number) with its RREQ message as described previously.

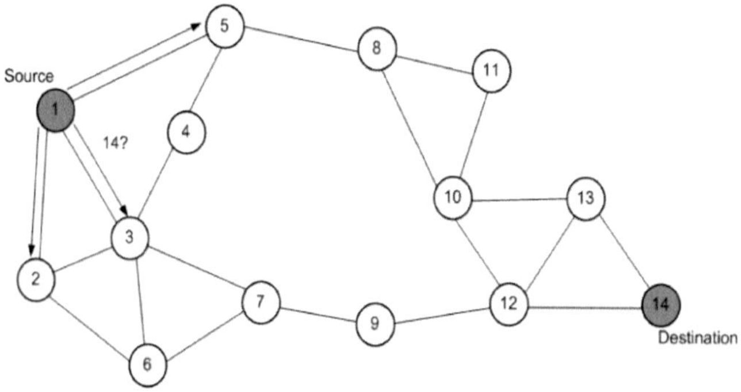

Figure 25: Broadcasting the RREQ Message [141]

On receiving the RREQ, the request receiving nodes forward it to the RREQ message source via a reverse route as shown in the next figure. This can be done by the use of a routing table. The entry of routing table is noted in the following figure. It can be noted that by receiving the request, the sequence number of the source node is compared to the DSN of the RREQ message. [145] have indicated that if the received RREQ message has a lower or equal sequence number, it is discarded. However, if the DSN is higher, the number of the current node will be updated with the new value of the sequence number.

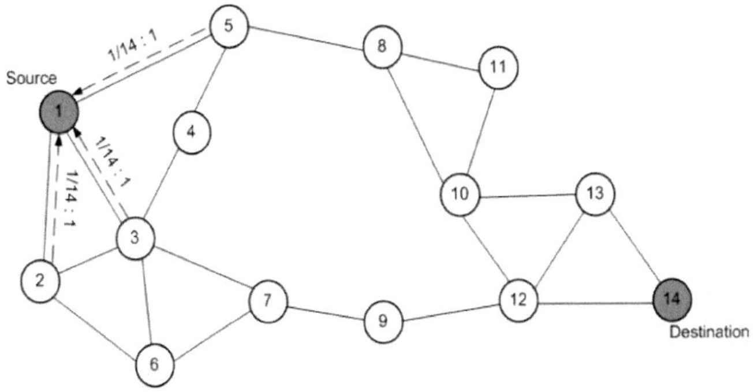

Figure 26: Forwarding a Reverse Route from the Receiving Nodes to the Source Node [141]

After that, node 1 floods the RREQ message again to the ad-hoc network. Then, the source neighbors (2, 3, and 5) send the request message to their neighbors and the hop count of the RREQ message increases with every hop. The main disadvantage of this promulgation is that the RREQ message will be sent to the originator too. In this situation, AODV protocol proposes that the nodes who have already received the request simply ignore and discard it due to the managed sequence number of the destination and the RREQ message

ID. The reason behind using this mechanism is due to the loop-freedom of this protocol [141]. This procedure has been illustrated in Figure 27 and 28 respectively.

It can be seen in Figure 29 that there is a continuous broadcasting of RREQ messages to the destination node. On the other hand, [146] have suggested that this continuous broadcasting can lead to several collision problems in MANETs. At the end, these messages reach the destination. As shown in Figure 30, there are two different detected routes.

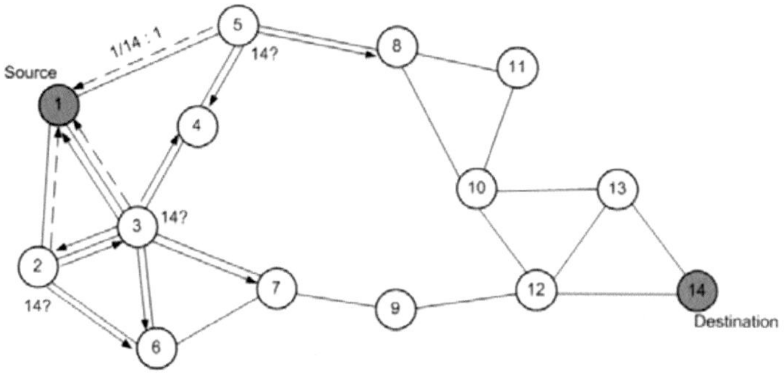

Figure 27: AODV Mechanism (1) [141]

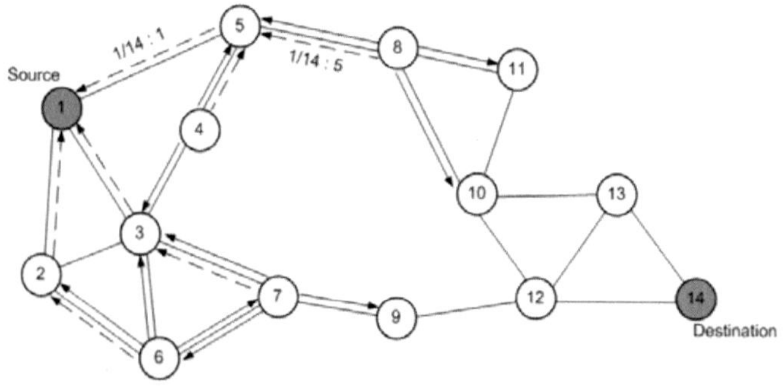

Figure 28: AODV Mechanism (2) [141]

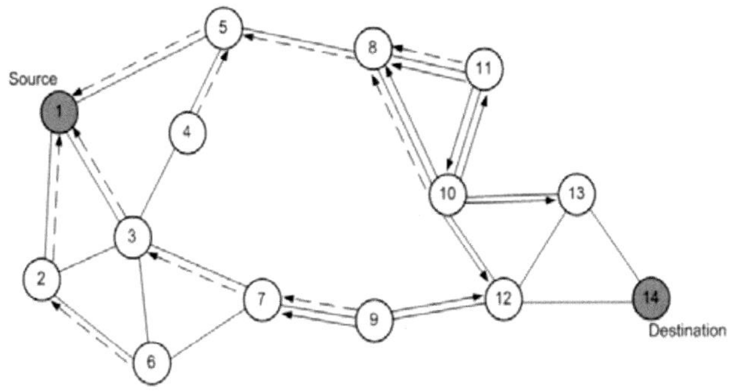

Figure 29: A Continuous Broadcasting of RREQ Messages to the Destination Node [141]

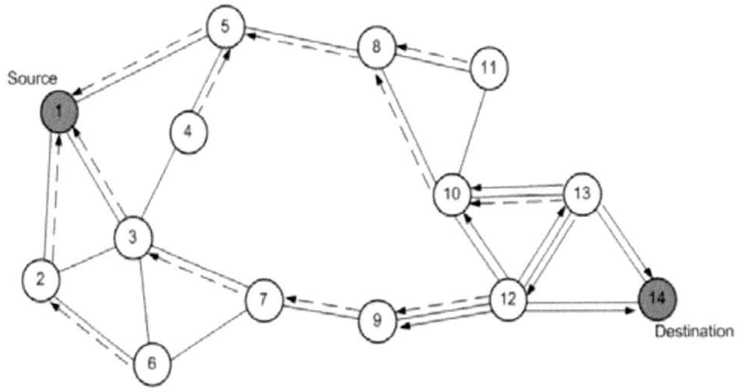

Figure 30: Illustration of the Received Messages by the Destination [141]

2.2.3.2 Route Reply – RREP

When the RREQ message is received by node 14, it will send back a RREP message to the sender node 1.

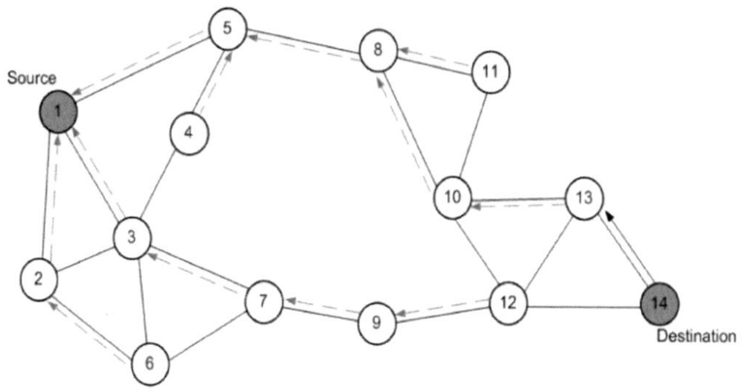

Figure 31: Illustration of Sending RREP Message from the Destination to the Sender Node [141]

Node 14 updates and increments its DSN before sending a RREP. After that, it is able to create the RREP to be sent to the destination. The destination sends a RREP as a unicast message and then it selects the shortest route back to the source. This can be done by comparing the hop count of the message [147]. The shortest path algorithm in the following figure is made randomly as both of these two routes have the same length. The selected path with the next hop in this case is node 13.

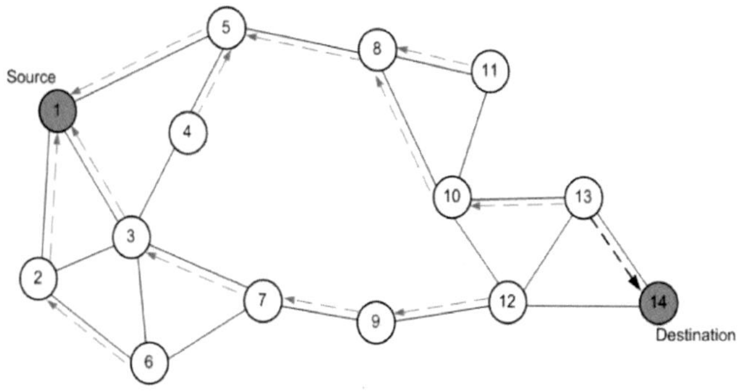

Figure 32: Selecting Node 13 as the Shortest Route [141]

In the next figure, a RREP message has been received by node 13, and it updates its routing table by checking the DSN again, then it can forward the RREP to the destination.

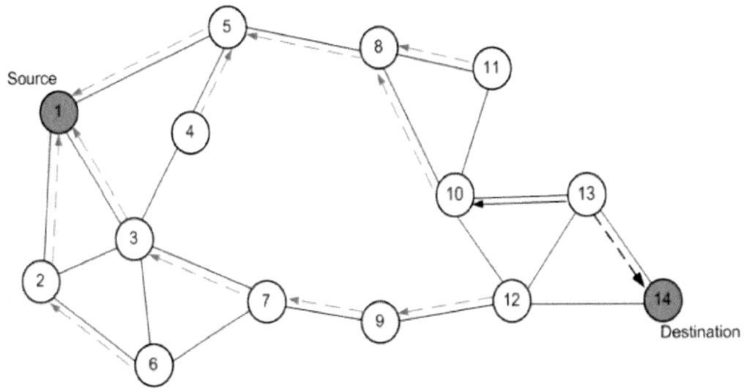

Figure 33: Receiving a RREP Message by Node 13 [141]

Thereby, the RREP message follows the reverse route of the RREQ message back to the source of the request.

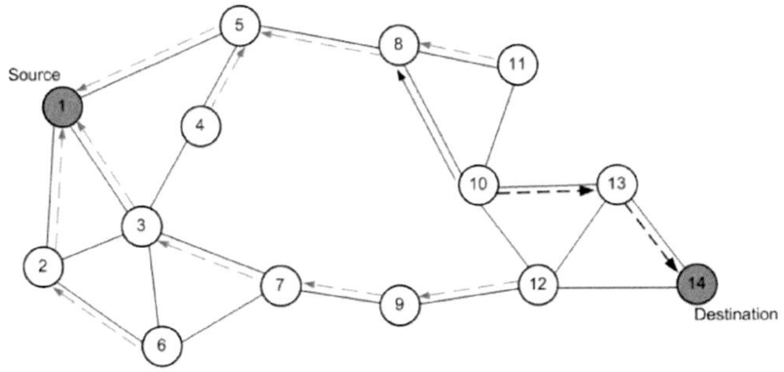

Figure 34: Illustration of the RREP Message Path by Following the RREQ Message Path [141]

After checking the DSN by each node in the ad hoc network, through which the RREP message passes, every node updates its routing table by adding a forward route to the destination into its table.

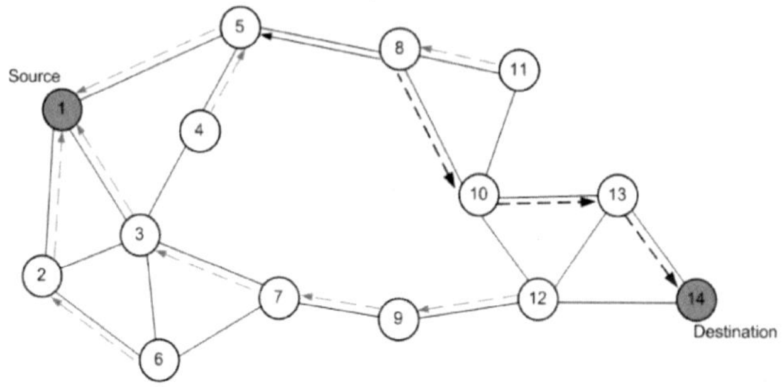

Figure 35: Adding Forward Route to the Destination by each Node into its Table [141]

It can be noted that the RREP message is much faster since a standard unicast message does not encounter the problems, which a broadcasting message could encounter while flooding through the ad hoc network [141]. The next two figures (36 and 37) show how the source node receives the RREP message and also how the forward route transfers the data to the destination node 14. It can be seen that when the source node receives the RREP message, it accepts automatically the current DSN of this message.

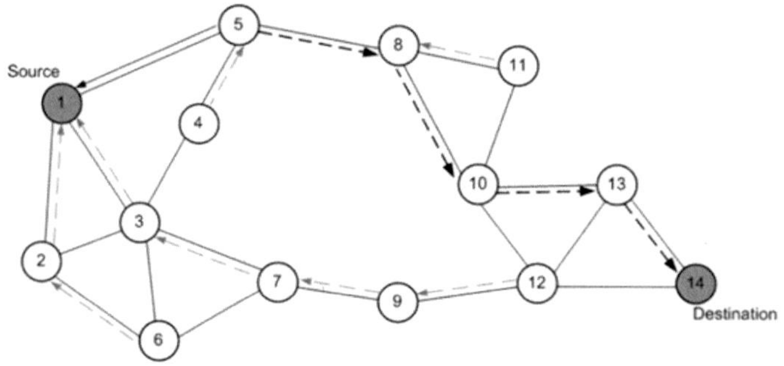

Figure 36: Illustration of Receiving the RREP Message by the Source Node [141]

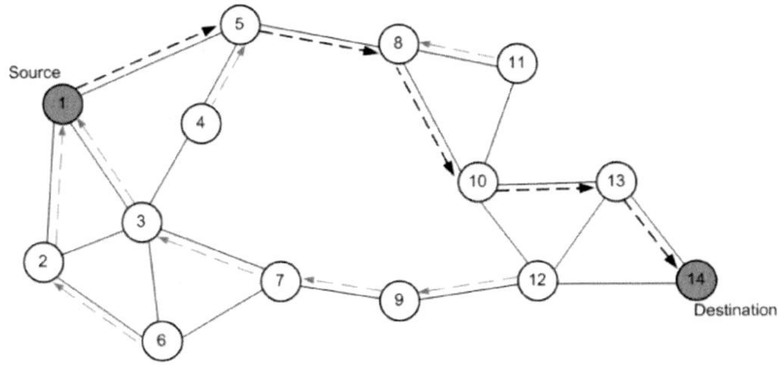

Figure 37: Illustration of Transferring the Data from the Forward Route to the Destination [141]

Finally, the route is created from node 1 to the destination route 14. After that, it can be clearly seen in Figure 38 that while the current route is still active, the data traffic can flow along the forward route. This forward route will be refreshed at each usage by the TTL and that of the reverse route simply times out.

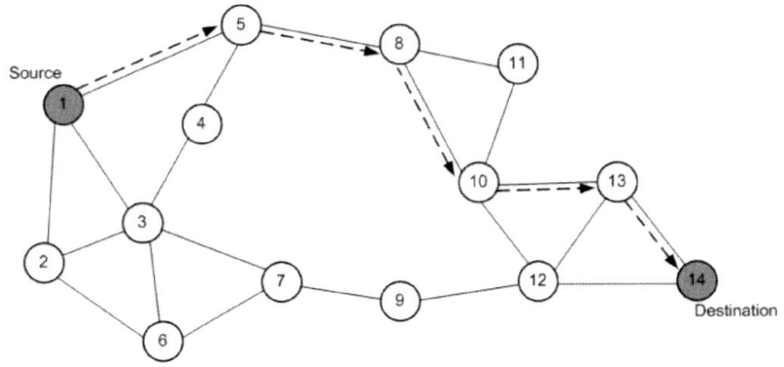

Figure 38: Illustration of the Data Path on the Forward Route [141]

2.2.3.3 Route Error – RERR

The route error detection and recovery is introduced to show if there is a broken link. It can be shown in Figure 39 that there is a broken link between nodes 8 and 10.

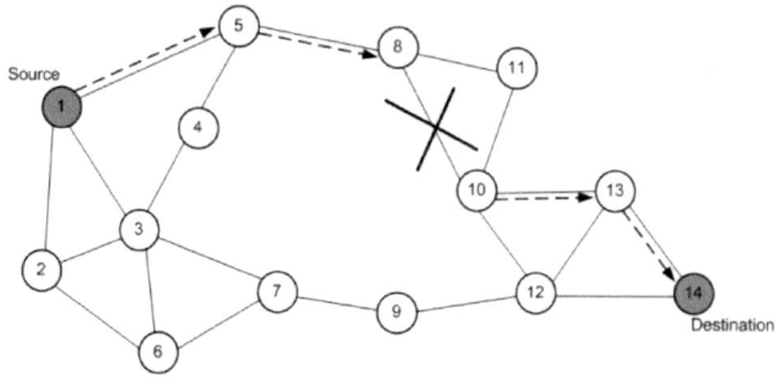

Figure 39: Illustration of the Broken Link between Nodes 8 and 10 [141]

Because of the traffic flow from source node 1 to destination node 14, node 8 will send data to node 10 after the link failure. Then, node 8 will be able to detect the broken link because the transmission of the data package to node 10 fails and this can be seen in the following figure.

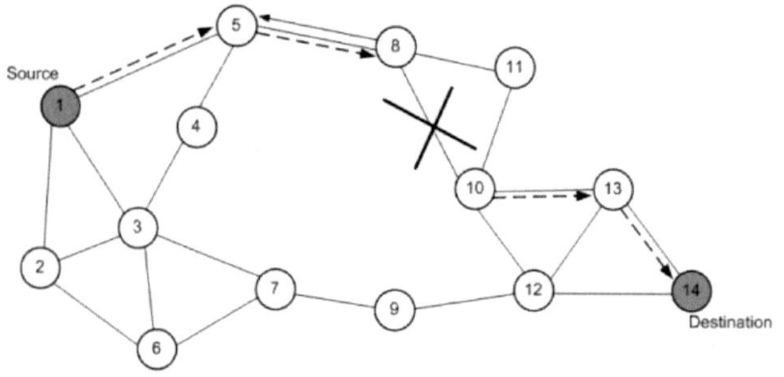

Figure 40: Illustration of Unsuccessful Data Transmission between Nodes 8 and 10 [141]

In terms of the failure detection, the sequence number will be incremented by node 8, and this node creates a RERR message of the broken route to the source node 1. Thus, there will be a new configuration for the RERR message as mentioned in the previous section.

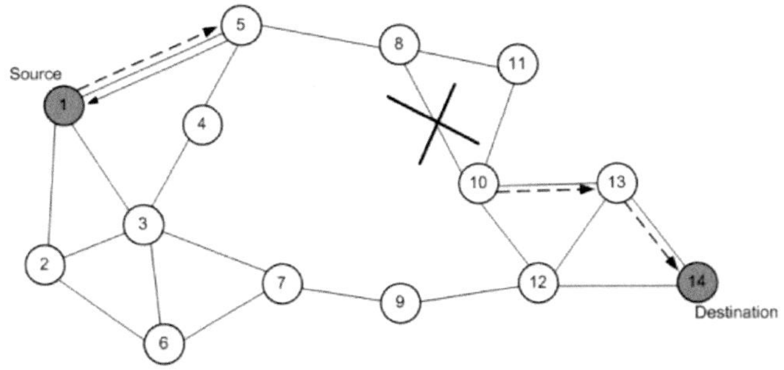

Figure 41: Illustration of Creating a RERR Message to Node1 [141]

In this case, RRER message returns to node 1 by using the reverse transmission route, until this message reaches source node 1.

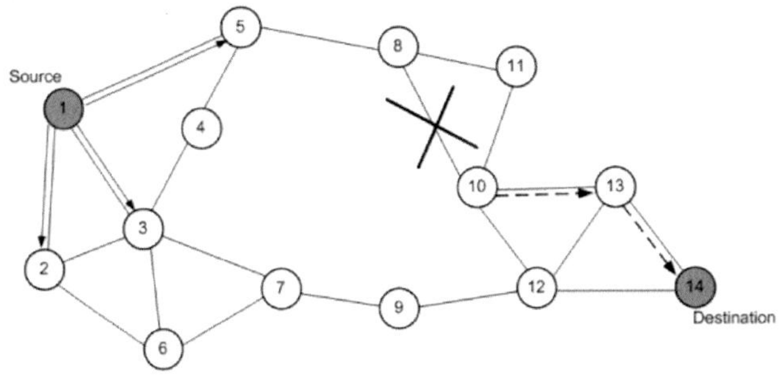

Figure 42: Illustration of Re-initiating the Route Discovery Process by Node1 [141]

Figure 42 shows how node 1 is required to re-initiate the route discovery process after receiving a RERR message by sending a new RREQ message. The description of the protocol process continues as described above. It is important to mention that node 10 or node 13 will send a RREP message back to the source when one of these nodes receives the RREQ. This happens when the node in its routing table has an active route to the destination.

2.3 Security Criteria of MANETs

MANETs were initially proposed to be used in dangerous environments, such as military and dangerous rescue operations [79][148]. Recent years have shown a tremendous increase in the usage of MANET in many fields as mentioned in the section on MANETs' applications, for example, military environments (automated battlefield), civilian environments, such as disaster recovery (flood, fire, earthquakes), and commercial, such as sensors on cars (car navigation safety). On this basis, it may be inferred that many users have become dependent on various wireless communication devices in their everyday activities. Hence, wireless networks have been an interesting area of research in the past few years and it is necessary for these networks to be efficient, reliable and secure.

In this section, many criteria are presented to review the existing approaches to secure MANETs. These criteria could be used to satisfy user requirements which involve packet drop ratio, average packet delay, and routing overhead. Within this section, each of these criteria is discussed in more detail.

2.3.1 Security Requirements

There are many measures, which are managed by a set of security policies, to address a specific feature of network security. Seven requirements have been identified in the security architecture to protect MANETs from all security threats [62][79][149] (see Figure 43):

- **Authentication** enables a node to ensure that it is communicated with the correct identity of peer node. The communicating nodes need verification of each other's identity to be satisfied that they are cooperating with the right party.

- **Authorization** protects network resources from unauthorized use and guarantees that only authorized nodes have access to the network.

- **Availability** means that assets have access at appropriate times to the authorized parties. This feature can be applied to data and services and it survives the network service despite denial of service attack.

- **Non-repudiation** ensures that sender or receiver nodes cannot disavow their sending or receiving of such a message. These nodes are liable for their actions and they are not allowed to deny the responsibility for their actions. This feature is important when there are disagreements or disputes over some events.

- **Data Confidentiality** means that the transmitted messages are secure from any unauthorized disclosure. This feature ensures that the data content cannot be intercepted by unauthorized entities and this can be done using any provided encryption techniques.

- **Data Integrity** means that messages cannot be modified during any communication but only the authorized users are able to alter them. These messages are protected from any unauthorized deletion, modification, replication and creation.

- **Privacy** protects the message information from the network activities observation and this can be achieved by protecting the node's identity and/or location in the network. This feature requires data encryption and more sophisticated techniques to hide the user's identity and/or location.

Figure 43 shows how these security requirements can be applied to the hierarchy of this network and it is important in order to provide security solutions for MANETs against all major security threats.

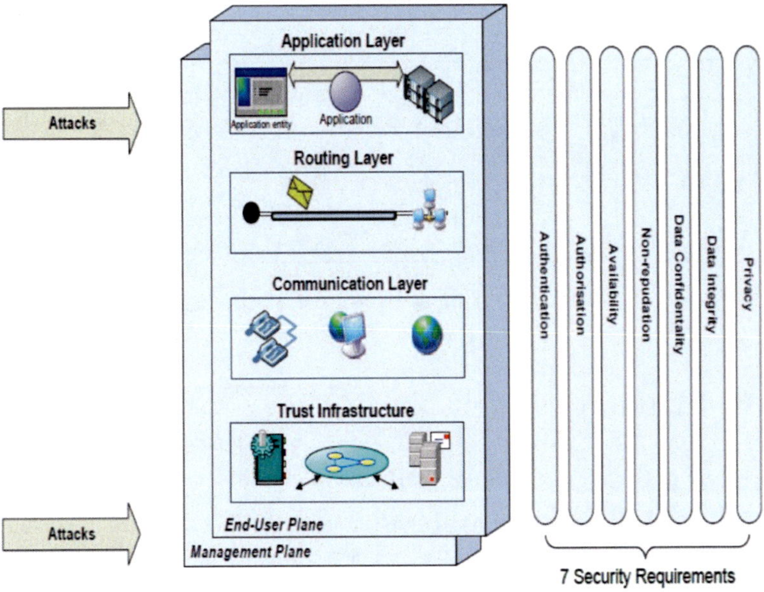

Figure 43: MANET Security Architecture [62]

2.3.2 MANET Attacks

In order to present some of the existing attacks on MANET, the different levels of security within this network are considered. In a standard network LAN or WLAN, there are 7 OSI layers, such as the Physical layer, Data link layer, Network layer, Transport layer, Session layer, Presentation layer, and Application layer. However, the security of a MANET can be divided into 5 OSI layers [77][150], such as the Application layer, Transport layer, Network layer, Data link layer and Physical layer. According to the specific layer, there are various types of attacks which differ in their essence. A classification list of these attacks can be seen in Figure 44. This section focuses on the Network layer attacks (Flooding attack and Black Hole attack).

MANet security layer	Attacks
Multi-layer attacks	DoS, impersonation, replay, MIMA
Application layer	Repudiation, data corruption
Transport layer	Session hijacking, SYN flooding
Network layer	Blackhole attack, Wormhole attack, Flooding attack,
	Colluding misrelay attack, Byzantine attack, Link Spoofing attack
Data link layer	Traffic monitoring and analysis,
	disruption MAC(802.11), WEP weakness
Physical layer	Jamming, interception, eavesdropping

Figure 44: Classification of Attacks in Different Layers [150]

2.3.2.1 Flooding Attack

The goal of a Flooding attack is to degrade the MANET's performance by flooding it with large amounts of traffic to disrupt the routing discovery or the maintenance phase within a MANET [35][150][151]. This can be done by a malicious node/ an attacker that aims not only to exhaust the network resources, such as network bandwidth, but also to consume the resources of an authentic network user, such as battery power. Furthermore, this kind of attack has an effect on the network performance by hindering the suitable execution of the routing algorithm (in the routing discovery phase) [77][150][152]. By using the AODV protocol in MANET, it can be noticed that an attacker sends multiple RREQs to non-existing destination nodes in a short time through RREQ flooding or routing table overflow. This means that the malicious node represents false routes to all authentic nodes within this network. The role of this malicious node is to prevent the creation of new actual ones; consequently, it causes routing table overflow by the authentic users. The avalanche of RREQs all over the network leads to the consumption of battery power and network bandwidth, causing DoS [77][150][152]. In

MANETs, every network participant (actual authentic user or simply node) is able to monitor and compute the evaluation of all neighbors' RREQs and this can be considered as a countermeasure against the Flooding attack. In the case of outmatching the RREQs' limit, which is preliminarily defined, the specific neighbor node comes with its ID in a blacklist. Then the authentic/actual node knows that it should not receive any RREQs from its neighbors which have been recorded in its blacklist. Moreover, this countermeasure can be enhanced when the RREQ limit is not preliminarily defined (or fixed), so the risk of attack with varying flooding rates is minimized [77][150].

2.3.2.2 Black Hole Attack

The aim of a Black Hole attack is to spuriously reply to any RREQs without having an active route to the precise destination, and to drop all the received packets [79] [150]. As in the Flooding attack, the Black Hole attack also concerns the AODV routing protocol in the network layer of MANET. It has been proposed that the process of a Black Hole attack has two steps [35][150][152]. The first step is that an attacker/ malicious node is able to modify the network topology by creating an auspicious "environment" for the attack. This can be done by advertising itself as having the shortest path for the packet to be delivered to the destination in order to intercept the packets between two authentic nodes. The second step of a Black Hole attack is that the malicious node depletes the interrupted data; this attack receives the information but it does not forward it to the destination node.

Figure 45 shows how the Black Hole attack works. In the beginning, the originator node sends RREQs to the network to discover valid routes. When the RREQ is received by the malicious node, sent by the source node, this node forwards it to the destination node and sends a RREP back to the source node as a legitimate route. After that, the source node transfers the data packets as an authentic user within the MANET. Then this malicious node intercepts the data flow by receiving the information but without forwarding it to the destination node. Obviously, the neighbor nodes are able to detect the sequence of the falsified RREQ or RREP messages, and then they put the malicious node in their blacklists by terminating the data flow over it [35][77][150][152]. It has been found that by minimizing the risk of being exposed, the malicious node cannot intercept the data transfer between two related nodes, but it is able to transmit the packets [150]. Furthermore, the attacker can adequately amend some messages sent from particular nodes, though not from all.

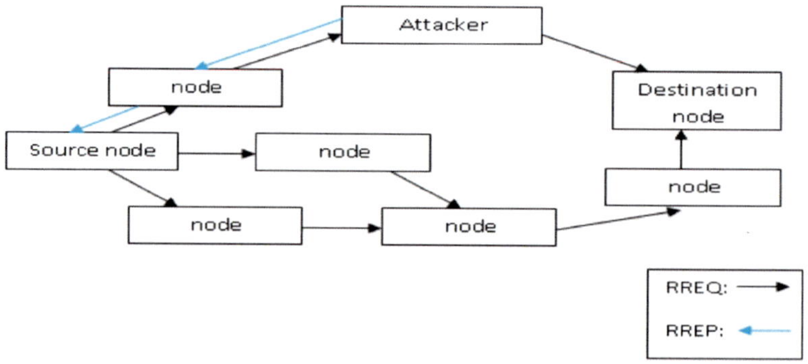

Figure 45: Black Hole Attack [150]

2.3.3 Performance Parameters

There are different kinds of parameters that have an effect on MANET performance. In this study, three parameters have been evaluated, not only to show the impact of Flooding and Black Hole attacks on the overall network performance, but also to show which attack is more dangerous to MANETs. These parameters are the packet drop ratio, average packet delay, and overhead routing. AODV routing protocol is required to examine these certain parameters. It is necessary to look at the control messages sent by the source by checking the effectiveness of this protocol to find a route to the destination. AODV routing protocol cannot be efficient if it has a longer delay compared to the other protocols that have shorter delays. However, this protocol can be efficient if it shows a high packet drop ratio compared to the other protocols that have low packet drop ratios. As a result, to select an efficient routing protocol, it is important to consider these selected parameters which have a great influence on any communication network.

2.3.2.1 Packet Drop Ratio

The PDR (***Packet drop ratio***) is the ratio of received data packets by the destination to the packets generated by the source [35][153]. It is represented in bits per second or packets per second. In MANETs, the packet drop ratio is affected by many changes in topology, limited bandwidth and limited power [154][155]. Unreliable communication is also one of the factors which adversely affects the success ratio [35][156]. Thus, it is important to improve the reliability of packet delivery to support many applications in different networking scenarios [156]. In most cases, the packet loss in MANET is greater than in wired networks or other infrastructure-based wireless networks [156]. This can be caused by a number of factors [156]:

- If the signal strength is very weak, the packets in MANET's wireless channels might be lost because of the noise and external interference [156][157]. Consequently, the SNR (***Signal to Noise Ratio***) of a signal in these networks can be several magnitudes smaller than those in wired networks. Furthermore, the intra-session interference contributes a non-negligible portion to the total interference because multiple wireless hops in MANETs share the same medium [158]. When the SNR becomes less than a certain threshold, transmission failures occur in the form of bit errors. If the error-correction mechanism of the communication system is not able to correct the bit errors, the whole data packet must be discarded [156][157]. Therefore, not only should a MANET's routing algorithm find short distance paths, but it should also find secure paths by considering the low signal power, mobility of nodes, and interference in these wireless networks [157].

- The wireless links in MANETs are opportunistic due to the mobility and power limitations. These links may disrupt the transmission of a data packet since the failures may happen at any time during the packet-transmission process. So, the packets loss will occur if there is no employed mechanism to discover and replace failed links [156].

- Network congestion can result in dropping packets. When the input buffer in a node is full, either a newly received data packet or a stored data packet has to be discarded to avoid buffer overflow [156].

In this research, the PDR has been used to identify the reliability of packet transmissions in MANETs since these networks are more vulnerable to various attacks [159].

2.3.2.2 Average Packet Delay

The delay refers to the time it takes to transmit a packet from its source to its destination, and this time is expressed in sec [160][161]. There are many applications, such as voice, which are sensitive to a packet delay and some of these applications require a low average delay in the network. Thus, a reliable transmission of packet data information, with low latency is essential in these wireless networks, employed in delay sensitive applications [162]. Packet average packet delay is a measure of routing protocol since the proper selection of a routing protocol to achieve maximum efficiency is a challenging task [160]. There are many different types of delays [160][155][163][164], such as:

- QD (***Queuing Delay***): The time the packet spends in routing queues.
- TD (***Transmission Delay***): The time it takes to push the packet's bits onto the link.
- PD (***Propagation delay***): The time for a signal to reach its destination.
- PD (***Processing delay***): The time routers take to process the packet header.

Mathematically, it can be shown as the following equation [155]:

$$d_{end-end} = N[d_{trans} + d_{prop} + d_{proc}]$$

Where

$d_{end-end}$ = End to end delay

d_{trans} = Transmission delay

d_{prop} = Propagating delay

d_{proc} = Processing delay

If there are n number of nodes, the total delay is the average of all the packets, network configuration, and source destination pairs [155][165].

It is important to mention that deploying large numbers of nodes uniformly facilitates the transmission of data packets with the guarantee of low-delay and low-error rate in the network [162][166]. It has been proposed that all of these delays depend on the efficiency of a processing algorithm and the computing power of the node [86][162]. According to the simulation results, average packet delay could be affected by a number of factors [167]:

- The delay increases with the growth of the number of nodes because of the channel contention and channel usage decay.

- Any arbitrary source-destination node pair can communicate with each other by a limited number of hops. Regarding the simulation results [167], by decreasing the transmission range of nodes in MANETs or decreasing the *LD (***Link Duration***), source-destination pairs are not able to connect with each other using the paths with a number of hops because this will lead to more delay.

- By increasing the average travel speed or decreasing the pause time (decreasing the LD), the mobility improves the instability of links [167]. This means that there is a broken link and rerouting becomes more frequent in the path, consequently packet delivery slows. Therefore, theoretically there should be a negative correlation between *LD* and delay.

LD (**Link Duration*****)** is the lifespan of a node-to-node link and this is calculated by the time the receiver enters the communication area to the time the receiver exits this [167][168].

2.3.2.3 Routing Overhead

The routing overhead is the number of routing packets transmitted per data packet delivered at the destination. Each hop-wise transmission of a routing packet is counted as one transmission packets [169][170]. It can be seen that the increase in the routing overhead reduces the performance of MANETs because it consumes the bandwidth which is available to transfer data between the nodes [170][171][172][173].

Routing Overhead = ***Total routing packets / Transmitted data packets.***

Routing overhead could be affected by:

- **The number of nodes:**

  The message overhead decreases as the number of nodes increases [174][175].

- **The degree of mobility:**

  The degree of mobility also has an impact on the overhead ratio. For instance, the overhead decreases by half when the node speed decreases by half. On the other hand, when there is an increase in the pause time, the overhead decreases only slightly. As a result, an increase in the node speed may lead to more topology changes and this causes more overhead to confront these changes [174].

2.3.4 Attack Prevention Techniques

This section gives an overview of Flooding attack prevention techniques since it has been shown from our results in chapter 5 that Flooding attack is more dangerous on MANETs than Black Hole attack and degrades the performance of it drastically. To defend MANETs against flooding attacks, a few of preventive security mechanisms have been proposed. FAP (***Flooding Attack Prevention***) and AMTT (***Avoiding Mistaken Transmission Table***) have been proposed to prevent Flooding attacks, since both of these techniques are the most significant schemes which prevent Flooding attacks.

2.3.4.1 FAP (Flooding Attack Prevention) Technique

FAP is a technique that has been proposed firstly by [61] in 2005. The definition of this technique as mentioned in is *"a generic defense against the Ad Hoc Flooding Attack in mobile ad hoc networks. When the intruder broadcasts exceeding packets of Route Request, the immediate neighbors of the intruder record the rate of Route Request. Once the threshold is exceeded, nodes deny any future request packets from the intruder"* [61].

2.3.4.1.1 FAP scheme

This scheme has two methods to resist the Flooding attack in Ad Hoc network:

- The procedure of neighbor suppression:

This method is used to prevent RREQ Flooding attack. Since the MANETs are multi-hop wireless networks, the node can send and receive packets from its neighbor nodes. If these neighbor nodes around the node refuse to receive its packets, this node cannot communicate with the others in MANET. By this way, the node has been isolated from the other nodes in the network [61].

The Figure 46 shows a simple topology of MANET. In this figure the node T attempts to communicate with the other nodes (R, K, and V) but these nodes refuse to receive packets from node T, so this node cannot forward any packet to the other nodes.

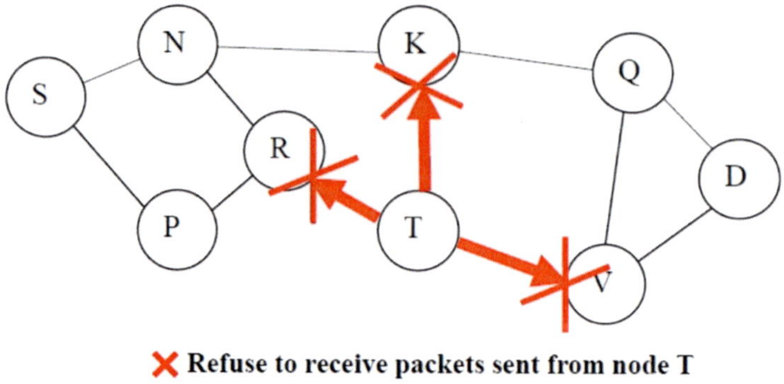

Figure 46: FAP in MANET [176]

- The procedure of path cut-off:

When the attacker originates this attack, the neighbor nodes have a difficulty to identify it and also to judge if the Data packets are useless in the network layer. In the application layer, the destination node can easily specify the received useless Data packets. The second scheme of FAP is Path Cutoff which is used to prevent Data Flooding attack [61].

When the attacker broadcasts Data Flooding attack, the attacker sets up a legal path to a victim node ahead. Then this attacker sends large number of futile data packets to the victim nodes. When these useless packets are detected by the nodes with FAP, they black list the attacker's path, which stops it to continue with the Flooding attack. The victim node broadcasts RERR messages to inform all the other nodes, that the attacker's IP address is unreachable [61]. Thus all the intermediate nodes, that receives the RERR, will remove the route from the attack to victim as shown in the next Figure.

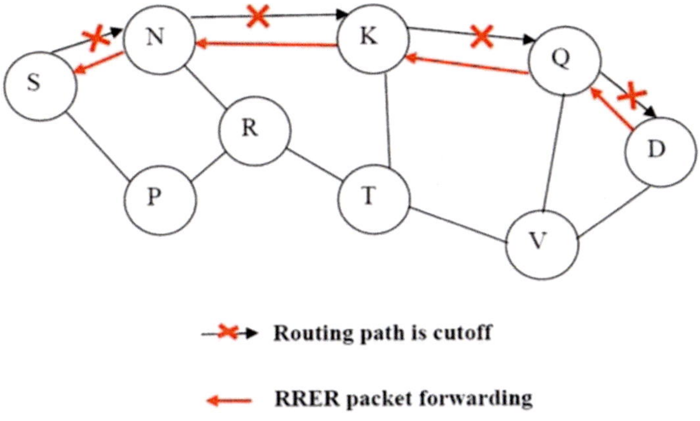

Figure 47: FAP in MANET [176]

2.3.4.1.2 Analysis on FAP Scheme

- When its heavy traffic in the network, this is important to compare the priority of each RREQ to select the effective forwarding order. Each node records the receiving RREQ and then calculate the transmission frequency for the neighboring nodes, as the frequency of sending RREQ depends on each RREQ's priority. The process of calculating the frequency is complicated since this process will burden the nodes in MANETs [61].

- FAP scheme employs passive defense against Data Flooding attack. If the attacker is able to set up routes with many valid nodes and is able to send concurrently lot of useless data packets, the effects to implement FAP are very severe [61].

- If more than one cooperative attacking nodes are present and somehow they are able to establish a valid path then they are able to send a huge amount of useless data, this will cause Data flooding attack in MANET. In this case, even the Path Cutoff is not effective because legal nodes cannot sense that both of sender and receiver are illegal nodes. So such attack cannot be resisted [61].

2.3.4.2 AMTT (Avoiding Mistaken Transmission Table) Technique

AMTT is a technique that has been proposed firstly in 2006. In this scheme as mentioned in [177], *"Legal nodes can distinguish illegal nodes and refuse to forward packages for them, so flooding attack can be defended"* [177].
This technique shows that it can resist flooding attack at a little cost more than FAP.

2.3.4.2.1 AMTT Scheme

- AMTT format

In this scheme, each node establishes an AMTT. The received RREQ packets are recorded in this table and they enroll the existed legal communication routes [177]. The following figure shows the format of AMTT.

S IP Addr	D IP Addr	RREQ Num	Seq Num	Vald Indic	Comm Rec

S IP Addr: the Source IP Address;

D IP Addr: the Destination IP Address;

RREQ Num: Number of RREQ Packets;

Seq Num: Sequence Number of RREQ;

Vald Indic: Validity Indication, 1 indicates this route is legal, NULL indicates it is illegal;

Comm Rec: Number of Data Packets Passed Through;

Figure 48: AMTT Scheme [177]

2.3.4.2.2 Analysis on AMMT Scheme

- The RREQ value is used by each node to record the number of RREQ packets forwarded by neighbor nodes. This node periodically statistics every RREQ Num in the table to detect whether a RREQ flooding attack is going on [61].

- In AMTT, the legal link is enrolled and cannot forward data packages for attacking nodes. Therefore, it can avoid data flooding in the origination [61].

2.3.4.3 The Comparison of FAP Scheme and AMTT Scheme

- The attack can be prevented by AMTT scheme and this can be done by the corporative work of two or more attacking nodes, and this cannot be solved by using FAP scheme [61].

- In AMTT scheme, the algorithm is simple because the table is dynamically constructed and deleted and it will not put much burden on each node [61].

- AMTT scheme has high precision in LANs but the security is not high and the traffic in each node is almost equal, such as many mobile hosts constructing a temporary conferencing LAN [61].

Chapter 3: Methodology

This chapter focuses on the research methodology and the technical tools used in this project. In this chapter the reasons behind choosing the network simulator NS-3 and also more details about this simulator can be found.

3.1 Research Methodology Categories

Research methodology describes how the development work should be carried out in the form of research activity. It is a tool that is used for collecting and analyzing the data and on the basis of this analysis conclusions are drawn. Research methodology is a combination of three categories, quantitative, qualitative and mixed approach [35][178].

3.1.1 Quantitative Approach

The Quantitative approach can be achieved by scrutinizing the problems collected from data, experiments and simulation results. These results are analyzed and then on the basis of these analyses, decisions are made. This approach is used when the researchers want to verify their proposed theories, or observe the information in greater detail [35][178].

3.1.2 Qualitative Approach

The Qualitative approach can be carried out by studying a collection *of* participatory and /or constructive perspectives. This approach includes different strategies, such as ethnographies, phenomenology, and grounded theories. The researchers use this approach when they want to study the context or focus on a single phenomenon or concept to achieve their desired goals [35][178].

3.1.3 Mixed Approach

A mixed approach includes both quantitative and qualitative approaches. The researchers use this approach to base their knowledge claims on practical grounds. This approach has the ability to produce the necessary complete knowledge (theory and practice) to combine both quantitative and qualitative approaches [35][178].

3.1.4 Author's Approach

Author's approach in this research has been conducted by using quantitative approach. This approach starts by studying the related work specific to security issues in MANETs and this is followed by simulating some of MANET scenarios. The results are collected and analyzed, and then the conclusions are drawn on the basis of results obtained from the simulation.

3.2 Research Design

The research conducted in this book was divided into four stages:

- Problem identification.
- Related work.
- Building simulation.
- Results analysis.

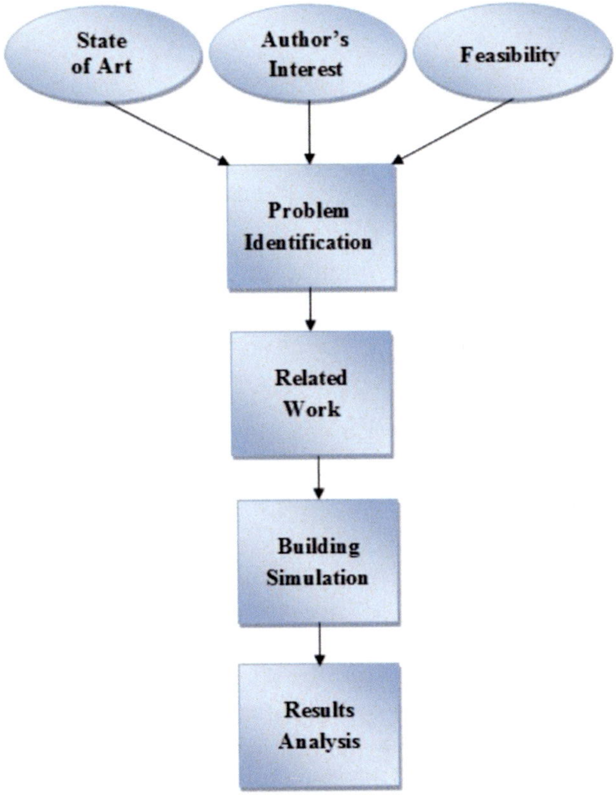

Figure 49: Research Methodology

- **Problem Identification:**

  The most important stage of this research is to select an appropriate problem area. Different problems have been investigated with the author's interest in mind. Much time has been devoted to this part to select a relevant and serious issue. The author selected the reliability of data dissemination in MANETs as the area of interest and within this network the focus was placed on the security issues.

- **Related Work:**

  When the problem was identified, the second stage was to review the state of the art. It is important to understand the capability of MANETs and the security issues involved in these networks. The related work was conducted to develop a solid background to this research. In addition, different security criteria of MANETs have been studied.

- **Building the Simulation:**

  Work-related knowledge is put together to develop and build a simulation. Different scenarios are developed according to the requirements of the identified problem and are then simulated.

- **Results Analysis:**

  One of the most important stages is analyzing the. The results obtained from the simulation are analyzed carefully and then, on the basis of this analysis, conclusions are drawn.

3.3 Simulation Tool

3.3.1 An Overview of the Simulation Tool

In this research, the simulation tool used for analysis is NS-3.14 released on 5 June 2012. This simulator is highly preferred for academic networking research since it demonstrated the best overall performance [179]. The definition of this simulator as mentioned on its website is:
"a discrete-event network simulator for Internet systems, targeted primarily for research and educational use. Ns-3 is free software, licensed under the GNU GPLv2 license, and is publicly available for research, development, and use" [180]. Figure 50 shows the modules of NS-3.

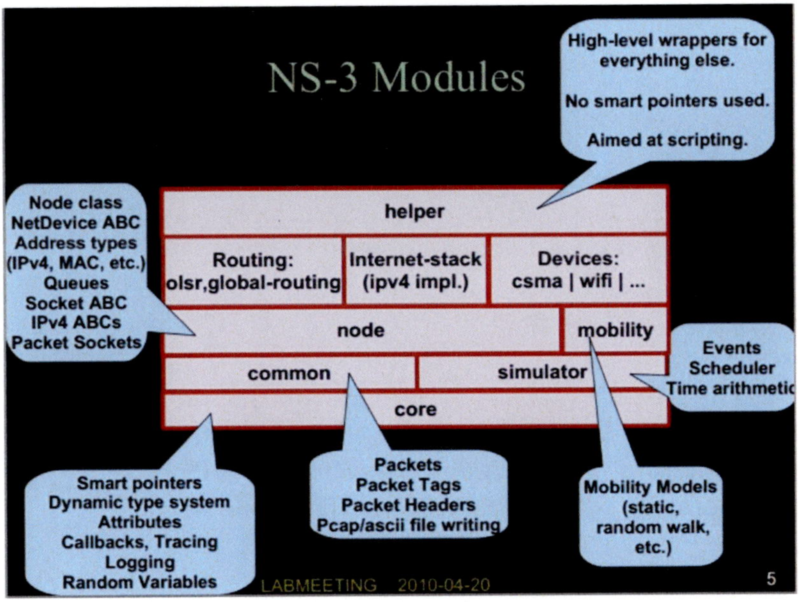

Figure 50: NS-3 Modules [181]

3.3.2 Features of NS-3

NS-3 [182] or Network simulator-3 will be used to simulate the AODV environment and different security attacks and prevention schemes for this project. Ns-3 is freeware software which was developed by University of California, Berkeley to simulate and monitor performance of existing or future data communication networks. NS-3 is open-source and licensed under GNU GPLv2, and anyone from academics, researchers and industry can contribute to it.

The NS-2 simulator [172] is the predecessor of NS-3 and was widely used; now certain improvements have been made to the ns-3 which is detailed below.

- **Software Core:**

  The software core is written in C++, which was also part of the ns-2 software; but the scripting interface has been changed to support Python scripting instead of OTcl. Several new additions such as call-backs, smart pointers and templates have been added to make it more modular and object oriented.

- **Realistic Modelling:**

  In ns-3 an attempt has been made to make the nodes closer to a real network and provide facilities which are present in actual computer nodes. This includes support of sockets, multiple interfaces and devices.

- **Integration:**

  Ns-3 is designed to adapt more open source software such as kernel protocol stacks, and routing demons.

- **Virtualization:**

  Support for lightweight virtual machines is added in ns-3. Posix-compliant applications can be easily ported now to run in simulation space.

- **Test Bed Integration:**

  Ns-3 has been enabled to support a real test bed. This enables the users to experiment with new ideas and protocols. This will also enable users to generate and consume network packets over real device drivers or VLANs.

- **Tracing:**

  Ns-3 has added a statistics gathering framework using call back-based design. This enables statistics output without rebuilding the simulation core.

Chapter 4: Description of the Work

MANETs are subjected to different types of security and service denial attacks. Due to the nature of the reactive routing protocols such as AODV, they are highly vulnerable to the DOS (Flooding) and Black Hole attacks. Although these two attacks are not limited to MANETs they have severe consequences on the performance of these networks as compared to fixed wireless or wired networks. The major reason contributing to the vulnerability of MANETs is the already scarce bandwidth, the shared medium (which causes long back off times) and battery life. In the previous chapters the nature of these attacks has been discussed in detail, in this current chapter the results of the experiments carried out in ns-3 and the discussion of these results are presented. In the next section, details of the topology of the simulation and the environment are presented followed by the presentation of the different simulation scenarios. In the later part of the chapter the results are presented both in graphical and tabular form and compared on the basis of three different network performance metrics. In the final section, a discussion and comparison of the findings of both the flooding and Black Hole impact is presented.

4.1 Topology

The simulation environment is based on a Wireless ad hoc network, this consist of 10 nodes. The nodes are users of Wi-Fi physical and MAC layer and the nodes move in a random walk based on the Gaussian Markov Mobility Model [183]. In this model, the velocity of the mobile node is assumed to be correlated over time and modelled as a Gauss-Markov stochastic process. The ns-3 code snippet for initializing the node is given below.

```
/************ Code snippet for initializing Random Gaussian
Markov mobility ***/

MobilityHelper mobility;

mobility.SetPositionAllocator ("ns3::RandomBoxPositionAllocator",
"X", RandomVariableValue (UniformVariable (0, 1000)),
"Y", RandomVariableValue (UniformVariable (0, 500)),
"Z", RandomVariableValue (UniformVariable (0, 500)));
mobility.SetMobilityModel ("ns3::GaussMarkovMobilityModel",
"Bounds", BoxValue (Box (0, 100, 0, 100, 0, 100)),
"TimeStep", TimeValue (Seconds (0.5)),
"Alpha", DoubleValue (0.95),
"MeanVelocity", RandomVariableValue (UniformVariable (0, 2)),
"MeanDirection", RandomVariableValue (UniformVariable (0,
6.283185307)),
"MeanPitch", RandomVariableValue (UniformVariable (0.05, 0.05)),
"NormalVelocity", RandomVariableValue (NormalVariable (0.0, 1.0,
10.0)),
"NormalDirection", RandomVariableValue (NormalVariable (0.0, 0.2,
0.4)),
"NormalPitch", RandomVariableValue (NormalVariable (0.0, 0.02,
0.04)));

mobility.Install (nodes);

/****************** Code snippet End
***********************/
```

The node is set to move with a mean velocity and direction based on a uniform distribution. This type of mobility model was chosen as it best describes the real world node movement in MANETs. A few of the mobility scenarios as captured by the NS3-netanim are given below:

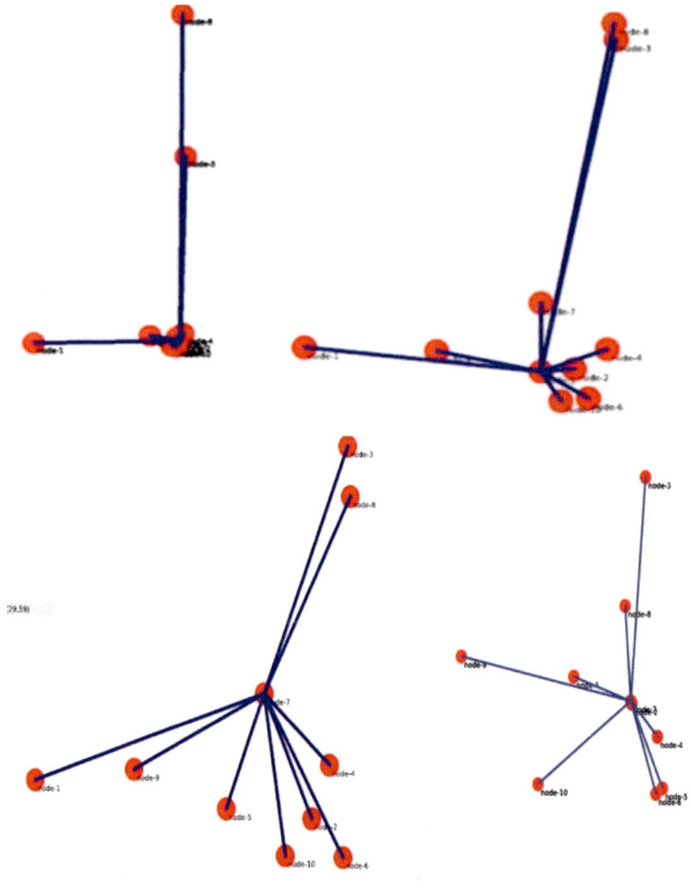

Figure 51: Snapshots of the Nodes Mobility (Markov Gaussian Random Walk) at 4 Different Instances of the Simulation

As described above the wireless nodes were initialized on a Wi-Fi physical and MAC layer, the Wi-Fi channel is modelled on a fading channel as implemented in NS-3 by YansWifiPhy helper. At the Internet layer IP routing in conjunction with AODV for wireless routing is used. The NS-3 has derived AODV as a sub-class of the Ipv4Routing main class, hence AODV inherits all the functions which are part of the Ipv4 routing and plus the extra methods and functions which are specific to the AODV protocol.

The Flooding and malicious nodes (Black Hole) nodes also used the similar models physical MAC and IP layers but we modified the application and routing behavior of these nodes to behave differently from the normal wireless nodes. The code is presented in Appendix A with comments. Here we present a brief overview of the Flooding node and malicious node implementation.

4.2 Flooding Attack

The Flooding node generates UDP traffic. To send the UDP packets to other nodes on the wireless network, the flooding node creates malicious sockets. Malicious sockets are created by spoofing the node's IP address and sending messages intended for unknown port numbers. When a node receives a packet which it is not bound to and listening then it wastes its processing time and it also generates a destination unknown ICMP packet, this wastes precious network and channel time. In this manner the flooding node keeps the wireless channel occupied with malicious information. This attack is specific to data packet flooding; another aspect of attack which is used in these scenarios is flooding of AODV hello messages. Normally AODV has a controlled number of hello packets which it can send in a given time so as to avoid replication and avoid wasting network bandwidth. A Flooding node violates this limit by flooding the network with an infinite number of hello packets. Although the AODV node tries to keep a check on the rate of hello packets sent by neighboring nodes, by not forwarding the hello messages from a flooding node, these flooding packets waste precious network bandwidth.

4.3 Black Hole Attack

Black Hole attack is implemented by identifying nodes to drop routing information received from neighboring nodes and also forwards wrong routing information. This is implemented to add an attribute "EnableMalicious" in the AODV routing class and this can be enabled using the Config::Set method as shown in the code snippet.
The attribute that is added in the AODV routing class is given below:

```
.AddAttribute ("EnableMalicious","Indicates whether a node is
malicious.",
               BooleanValue(false),
               MakeBooleanAccessor (&RoutingProtocol::m_ismalicious),
               MakeBooleanChecker ())
```

This attribute is set in the main function using this code.

```
if(malicious_enable)
{
Config::Set("/NodeList/[malicious_node]/$ns3::aodv::RoutingProtocol/
EnableMalicious",BooleanValue(true));
```

4.4 Network Performance

To compare the impact of flooding effects we first analyzed the normal network without any type of flooding nodes. We used three different parameters to determine the performance of the network. These parameters are packet drop ratio, average packet delay and routing overhead.

4.4.1 Packet Drop Ratio

This is the ratio of the total number of transmitted packets to the total number of received packets; this is expressed in terms of % so the resultant number is then multiplied by 100.

4.4.2 Average Packet Delay

This value is attained by total transmission time divided by the number of packets received. This value is attained by using DelayJitterEstimation class in NS-3. A time tag is attached to each packet transmitted and when it reaches the destination this information is subtracted from the simulator::Now function which keeps track of the current time. The Code snippet shows the implementation summary.

```
/**** Code at the transmitter side *****/

DelayJitterEstimation delay;

delay.PrepareTx(p); // p is the data packet to be transmitted

/**************** code end *******************/

/********* Code at the receiver end **************/

delay.RecordRx(packet);
total_delay += delay.GetLastDelay().GetSeconds();//sum of packet
delays

/********** code end *******************/
```

4.4.3 Routing Overhead

This is the ratio of the number of routing bytes transferred on the network to the total amount of information travelled on the network. This total amount of information on the network includes the valid data bytes added to the routing information. An increase in routing information is inversely proportional to the useful throughput of the network. The Routing overhead is calculated by using a packet monitoring software called wireshark [2]. Wireshark uses either online packet transmissions to capture data or can extract information from files stored in .pcap format. Ns-3 supports the class to enable packet capture.

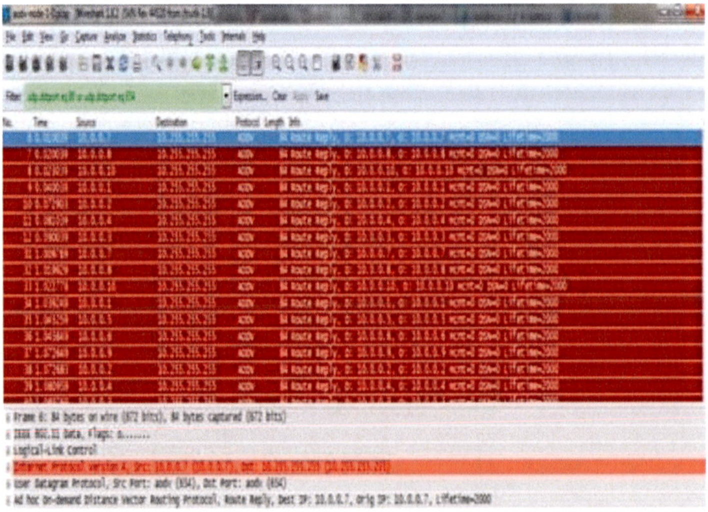

Figure 52: Wireshark Packet Capture

The function used to capture packets is given as

```
if (pcap)
  {
   wifiPhy.EnablePcapAll (std::string ("aodv"));
  }
```

The pcap file is generated at each node then using the mergecap tool in linux we merged all the node files except the flooding/malicious node. The wireshark provides filters to segregate irrelevant packets and then analyze the statistics. In our case since the valid data packets we were sending belonged to port 80 and the AODV routing packets has a unique port number 654, we filtered these packets from the flooding packets. This is done by using the filter *"udp.dstport eq 80 or udp.dstport eq 654"*. Then, the total packet statistics were analyzed as shown in the snapshot.

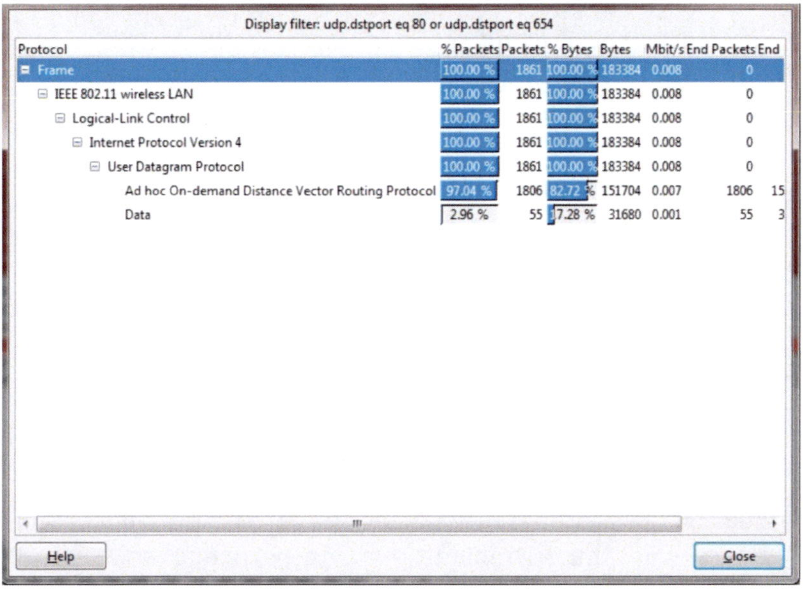

Display filter: udp.dstport eq 80 or udp.dstport eq 654

Protocol	% Packets	Packets	% Bytes	Bytes	Mbit/s	End Packets	End
Frame	100.00 %	1861	100.00 %	183384	0.008	0	
IEEE 802.11 wireless LAN	100.00 %	1861	100.00 %	183384	0.008	0	
Logical-Link Control	100.00 %	1861	100.00 %	183384	0.008	0	
Internet Protocol Version 4	100.00 %	1861	100.00 %	183384	0.008	0	
User Datagram Protocol	100.00 %	1861	100.00 %	183384	0.008	0	
Ad hoc On-demand Distance Vector Routing Protocol	97.04 %	1806	82.72 %	151704	0.007	1806	15
Data	2.96 %	55	17.28 %	31680	0.001	55	3

Figure 53: Wireshark Statistics Snapshot

In the snapshot, we can clearly see the number of data packets and AODV routing packets in terms of percentages of the total information exchanged on the network.

4.5 Scenarios

For the Flooding attack, we set up three different scenarios. These scenario were chosen to see the effects of flooding node positioning, the type of flooding and valid data packets.

Before discussing these scenarios we define three different types of nodes in the simulation.
Nodes can have three different roles in these simulation studies:

- **Message Transmitting Nodes:**

  Wireless nodes sending UDP packets on a user define a port either sending to a particular host or broadcasting to all the nodes on the network.

- **Message Receiver Nodes:**

  These nodes were only listening and receiving the messages on a well-known port, i.e. 80.

- **Flooding Nodes:**

  These are similar to the AODV wireless nodes with a difference that they flood UDP packets on MANET and the destination port was unknown. The flooding nodes were either transmitting to a particular node (for instance, unicast) or broadcasting to all the nodes.

The values chosen for these parameters in running the experiments are:

The inter-packet interval of flooding messages was varied between 1.2 to 0.1 seconds. The reason behind this variety (from the biggest to the lowest number) is to show that the largest interval has less flooding packets than the lowest interval which has more flooding packets.

The three scenarios are as follows:

- Scenario 1:

All the valid data packets and the Flooding packets were broadcasted on the network, this was used to analyze the effect of flooding on nodes which work as central or relay nodes.

- Scenario 2:

All the flooding packets were broadcasted but the valid data messages were unicast. This was used as a measure to find the effect of the Flooding on the one to one communication.

- Scenario 3:

The third scenario was setup with both the flooding node and the data transmission nodes sending the messages to a similar node. This was done to analyze the impact of flooding on the processing times of a receiver node which is receiving valid as well as malicious messages.

Malicious Node Scenario:

To analyze the behavior of a malicious node we ran experiments by assigning each node as malicious and then analyzing different parameters. In the Black Hole attack valid routing packets are dropped and malicious routing packets are generated hence the position of the node in the network is very important. This is why we chose differently located nodes with various mobility directions and analyzed the impact on the network performance.

Chapter 5: Evaluation and Discussion

This chapter highlights on three parts:

- The first part aims to explore which attack (Flooding attack or Black Hole attack) has more impact on MANETs and how it seriously degrades its performance.

- The second part aims to find the best prevention techniques for the most dangerous type of attack.

- The third part is the conclusion and provides suggestions for future work.

5.1 Network Performance under Attack

In this section, we will analyze which attack (Flooding or Black Hole attacks) has more impact on MANETs and how they dramatically degrade its performance. In the next section, we will simulate and find the best prevention technique for use with the most dangerous attack.

The experimental results of this research are collected from the simulator NS-3. This simulator provides a report that contains a summary of all the operations done during the execution time for a specific setting scenario. After that, these results are converted to diagrams to show the differences easily among several scenarios. In addition, these results will be evaluated and discussed in terms of packet drop ratio, average delay, and routing overhead. These metrics will be individually argued for each experimental scenario to compare the scenarios.

5.1.1 The Impact of Flooding Attack

This section shows the experimental results of Flooding attack under three different scenarios as discussed before:

5.1.1.1 Scenario 1

Flooding Broadcast
Message Packet Broadcast

Table 1: Network Performance Statistics for Flooding Scenario 1

Flooding Interval (sec)	No. of Sent Packets	No. of Received packets	Packet Drop Ratio (%)	Time Delay	Average Packet delay (sec)	Routing overhead (%)
1.2	137	1024	6.569343066	0.826903	0.000807522	50.02
1.1	137	1015	7.390510949	0.813164	0.000801147	50.12
1	137	1007	8.120437956	0.80283	0.000797249	50.55
0.9	137	1013	7.572992701	0.833178	0.000822486	50.61
0.8	137	994	9.306569343	0.81433	0.000819245	50.43
0.7	137	978	10.76642336	0.79849	0.000816452	50.71
0.6	137	970	11.49635036	0.80915	0.000834175	50.82
0.5	137	931	15.05474453	0.7665	0.000823308	51.71
0.4	137	890	18.79562044	0.734	0.000824719	52.88
0.3	137	836	23.72262774	0.72091	0.000862333	54.52
0.2	137	691	36.95255474	0.621043	0.00089876	58.34
0.1	137	325	70.34671533	0.436083	0.001341794	69.46

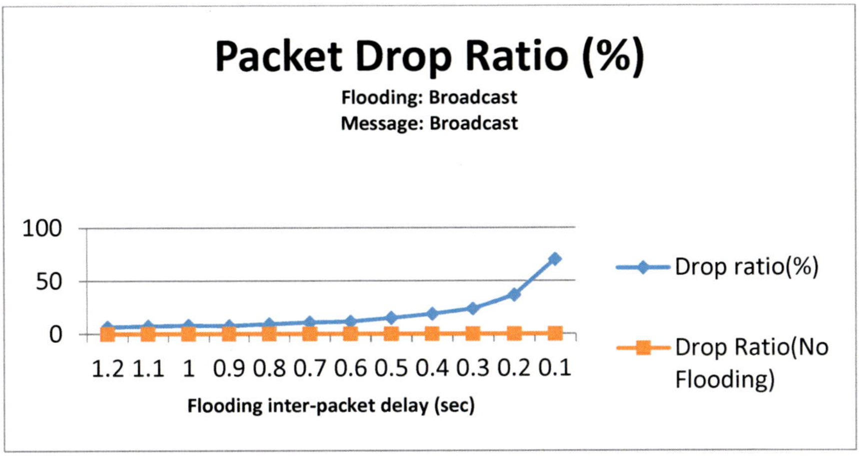

Figure 54: Comparison of Packet Drop Ratio for Scenario1

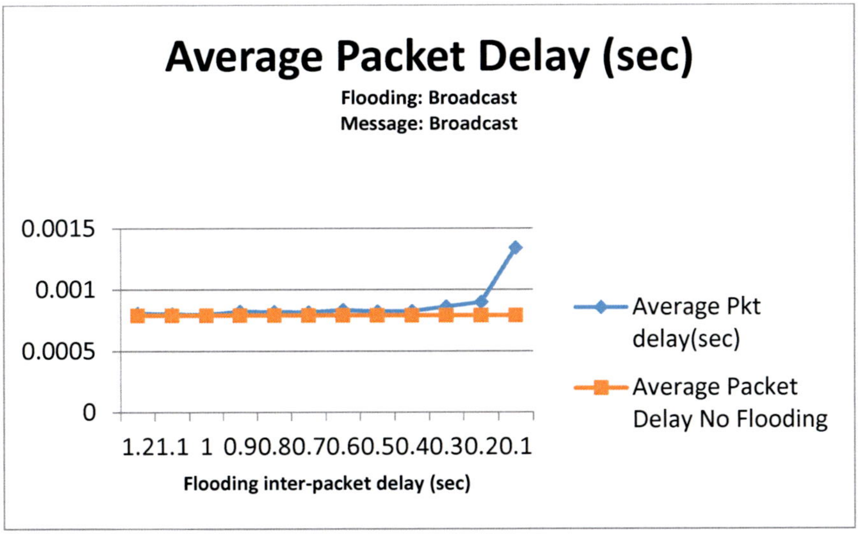

Figure 55: Comparison of Average Packet Delay for Scenario1

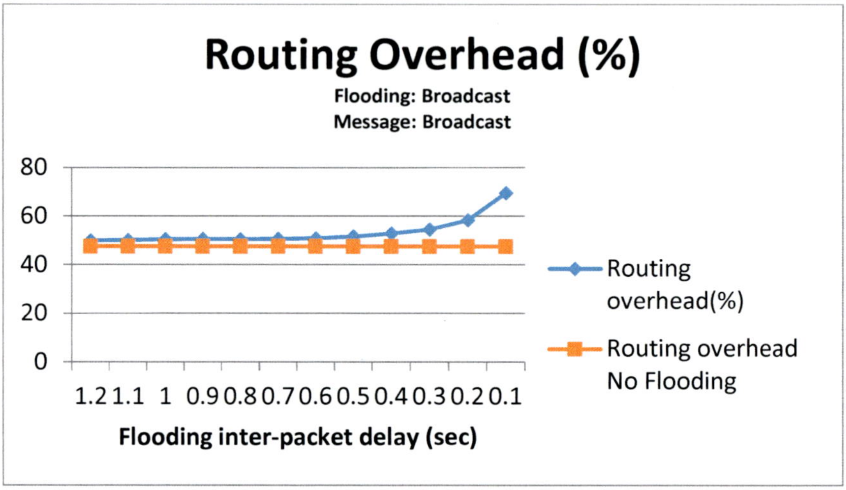

Figure 56: Comparison of Routing Overhead for Scenario1

Scenario 1:

The Flooding node and Message transmitting node are both broadcasting messages.

The results showed the following trend:

- The Packet Drop Ratio is 0 for no flooding but increased from approximately 6% to 80% for high rate of flooding.
- The Average Packet Delay also showed a dramatic increase for high flooding packet transmission rates from 0.3 sec. to 0.1 sec.

- Routing overhead showed an increase from 3% to 4% for low frequency of flooding messages as compared with no flooding scenario but increased significantly to approximately 23% for flooding message frequency of 0.1 sec.

Summary of Scenario 1:

This scenario clearly indicates that the network performance has seriously deteriorated with the increase of flooding packets. This impact is due to large packet processing times on the receiver nodes and useless routing information exchange.

5.1.1.2 Scenario 2

Flooding Broadcast
Message Packet Unicast

Table 2: Network Performance Statistics for Flooding Scenario 2

Flooding Interval (sec)	No. of Sent Packets	No. of Received packets	Packet Drop Ratio (%)	Time Delay	Average Packet delay (sec)	Routing overhead (%)
1.2	137	137	0	0.145541	0.001062343	65.72
1.1	137	137	0	0.148716	0.001085518	65.74
1	137	137	0	0.1515	0.001105839	65.8
0.9	137	137	0	0.1571	0.001146715	65.83
0.8	137	137	0	0.158772	0.00115892	65.88
0.7	137	137	0	0.1665	0.001215328	65.9
0.6	137	137	0	0.165773	0.001210022	66.1
0.5	137	137	0	0.181029	0.00132138	66.23
0.4	137	137	0	0.188691	0.001377307	66.33
0.3	137	137	0	0.209016	0.001525664	67.78
0.2	137	93	32.11678832	0.177956	0.001913505	74
0.1	137	0	100	infinite	0.005	100

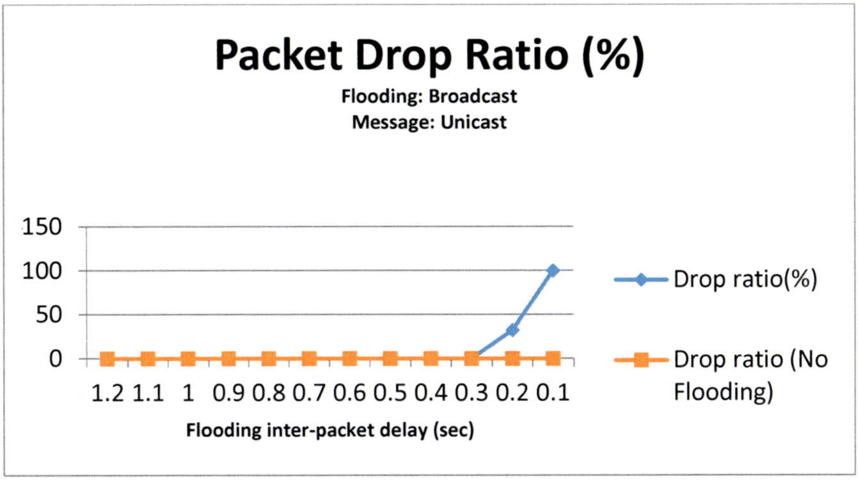

Figure 57: Comparison of Packet Drop Ratio for Scenario2

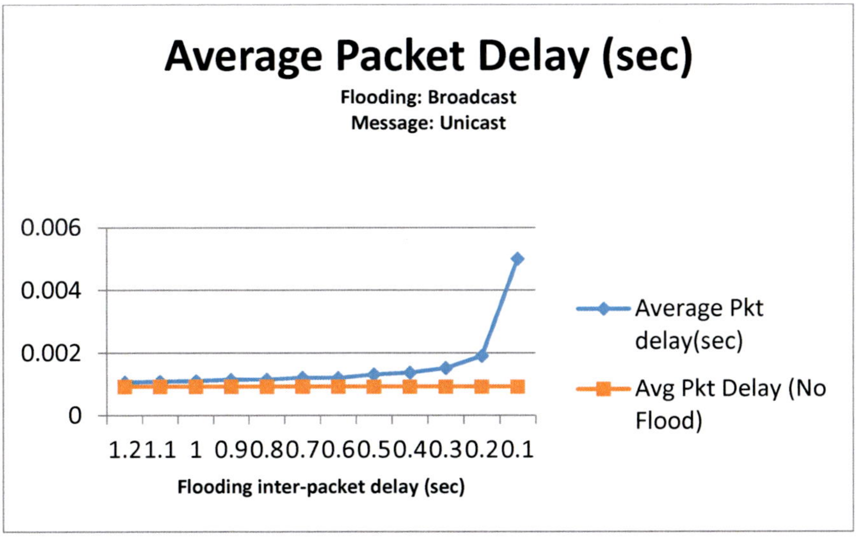

Figure 58: Comparison of Average Packet Delay for Scenario2

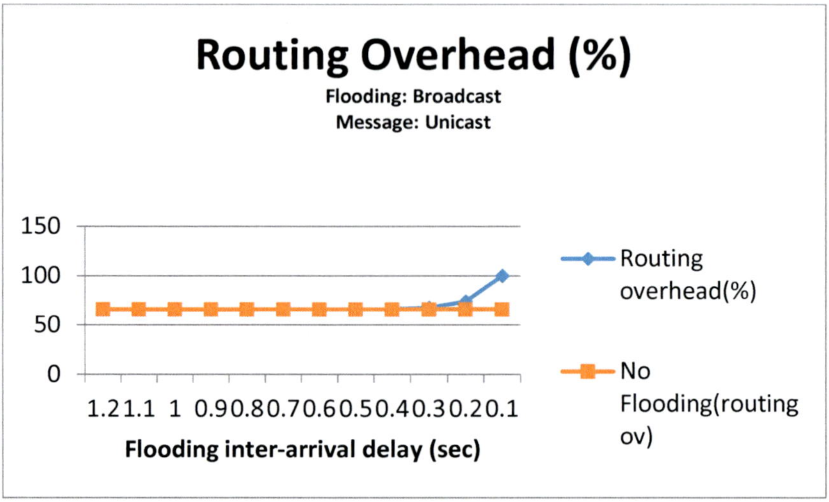

Figure 59: Comparison of Routing Overhead for Scenario2

Scenario 2:

The Flooding node is broadcasting while Message transmission is unicast.

The results showed the following trends:

- Routing overhead was same for the non-flooding and low frequency flooding scenario but increased to almost 100% for high flooding.
- Packet drop ratio is almost 0 for flooding frequencies of 0.4 sec but showed sudden increase for a high frequency of flooding.

- Average Packet delay also showed a drastic increase for high flooding packet transmission rates from 0.3 to 0.1 sec. The average packet delay is around 1ms for no flooding scenario while for high flooding the message delay increased till 5ms.

Summary of Scenario 2:

Again the statistics show that Flooding has an impact on the network performance, so much so that for large flooding rates the packet drop ratio is 100%. This means that the channel is fully occupied by the flooding packets and faulty routing information. The node for the valid transmission has no chance to transmit the packets. In the case of low flooding the transmission is unicast, therefore, there are more chances for the unicast packets to reach the destination due to the large interval of flooding packets. It is interesting to note that the routing overhead in this scenario is quite high, as in scenario 1, showing the fact that the number of routing packets exchanged per actual information transmitted is quite high.

5.1.1.3 Scenario 3

Flooding Unicast
Message Packet Unicast

Table 3: Network Performance Statistics for Flooding Scenario 3

Flooding Interval (sec)	No. of Sent Packets	No. of Received Packets	Packet Drop Ratio (%)	Time Delay	Average Packet Delay (sec)	Routing Overhead (%)
1.2	137	137	0	0.137996	0.00100727	65.95
1.1	137	137	0	0.13954	0.00101854	65.9
1	137	137	0	0.14171	0.00103438	65.9
0.9	137	137	0	0.13964	0.00101927	65.9
0.8	137	137	0	0.1464	0.001068613	65.9
0.7	137	137	0	0.141733	0.001034547	65.9
0.6	137	137	0	0.147803	0.001078854	65.9
0.5	137	137	0	0.14738	0.001075766	65.92
0.4	137	137	0	0.158742	0.001158701	65.98
0.3	137	137	0	0.166189	0.001213058	66.12
0.2	137	93	32.11678832	0.127931	0.001375602	73.96
0.1	137	0	100	0	0.005	100

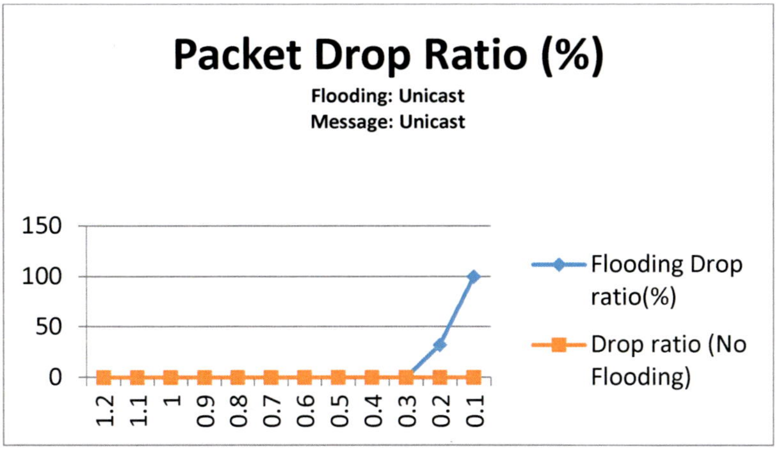

Figure 60: Comparison of Packet Drop Ratio for Scenario3

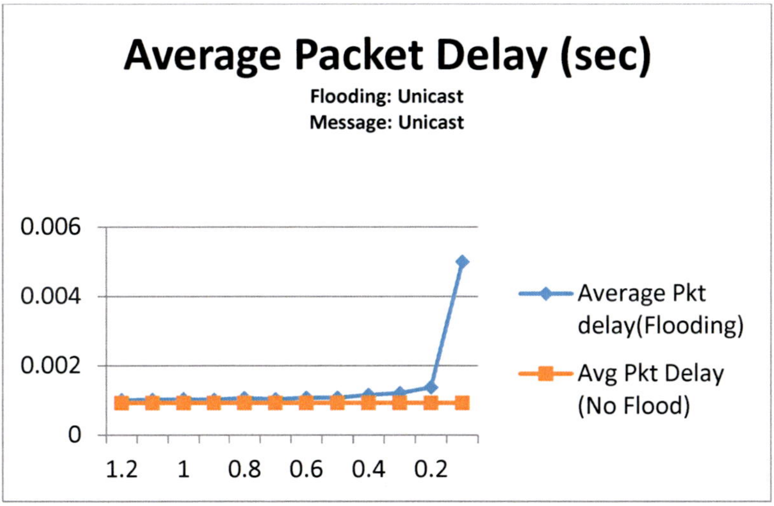

Figure 61: Comparison of Average Packet Delay for Scenario3

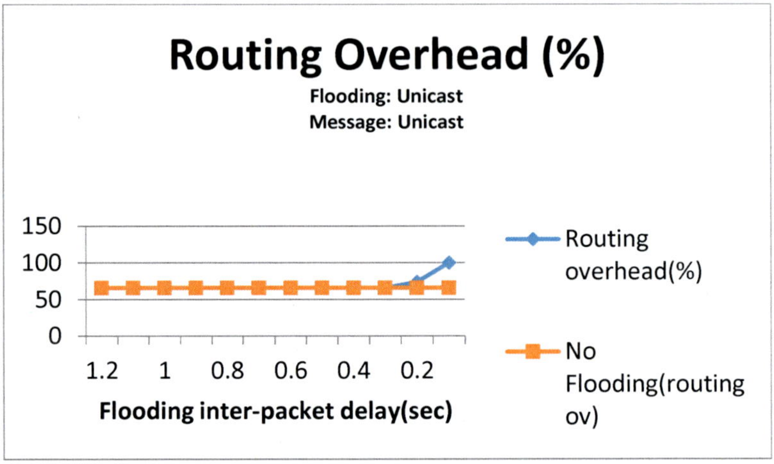

Figure 62: Comparison of Routing Overhead for Scenario3

Scenario 3:

The Flooding node is unicasting messages to the same receiver node as the valid message transmitting node.

The results showed the following trend:

- All the three parameter values were almost same as for no flooding scenarios to the low flooding values. Although there was a sudden increased observed for higher rates of flooding.

5.1.1.4 Summary of Flooding Attack

The high rate of flooding will degrade the performance of the network further. The wireless medium is occupied mainly by the flooding messages and also the nodes spent lot of their processing time in handling unknown messages.

Normally a message received for an unknown port generates an ICMP destination unreachable message and as more of these flooded messages arrive, the more the network performance is degraded.

5.1.2 The Impact of Black Hole Attack:

Statistics of the Network performance are given below when the wireless network is under Black Hole attack. Node 1 and 10 are the Transmitting and Receiving nodes.

Table 4: Statistics Obtained after Black Hole Attack

Malicious Node	No. of Sent Packets	No. of Received packets	Packet Drop Ratio (%)	Time Delay	Average Packet delay (sec)	Routing overhead (%)
2	137	813	25.82116788	0.644059	0.0007922	51.22
3	137	685	37.5	0.5426	0.000792117	52.15
4	137	1086	0.912408759	0.860324	0.000792195	48.73
5	137	684	37.59124088	0.5434	0.000794444	52.43
6	137	1086	0.912408759	0.860324	0.000792195	48.73
7	137	821	25.09124088	0.6521	0.000794275	50.82
8	137	801	26.91605839	0.659	0.000822722	51.15
9	137	950	13.32116788	0.75251	0.000792116	51.32

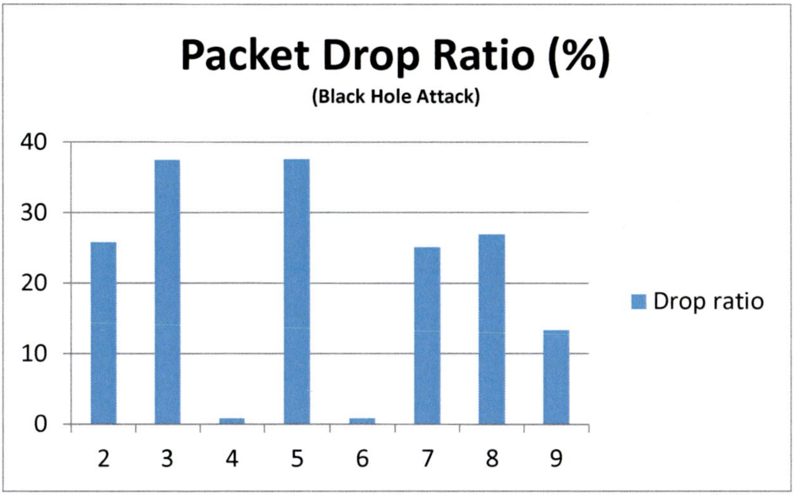

Figure 63: Packet Drop Ratio for Black Hole Attack

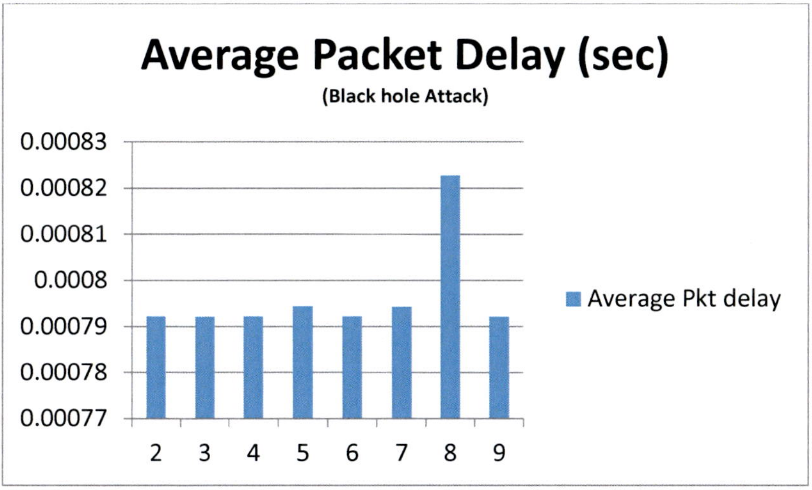

Figure 64: Average Packet Delay under Black Hole Attack

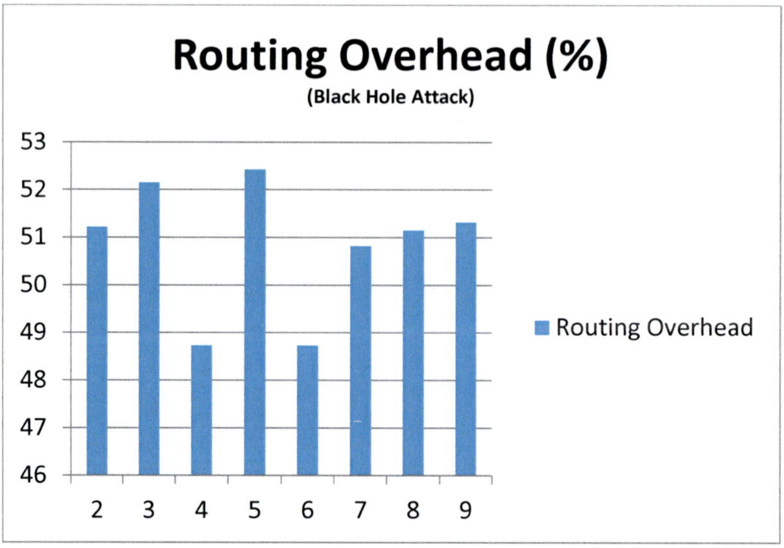

Figure 65: Routing Overhead under Black Hole Attack

Summary of Black Hole Attack

Based on the location of different nodes, the Black Hole attack has quite a variable impact on the network performance. As can be seen, the packet drop ratio varies between 0% and 40 %. This shows a large variance; although the average packet delay and the routing does not show such a large variance.

5.1.3 Comparison between Flooding and Black Hole Attacks

The network performance under three different flooding scenarios and Black Hole attack was presented in the last section. Now we will compare their effects and determine which attack more severely degrades the network's performance.

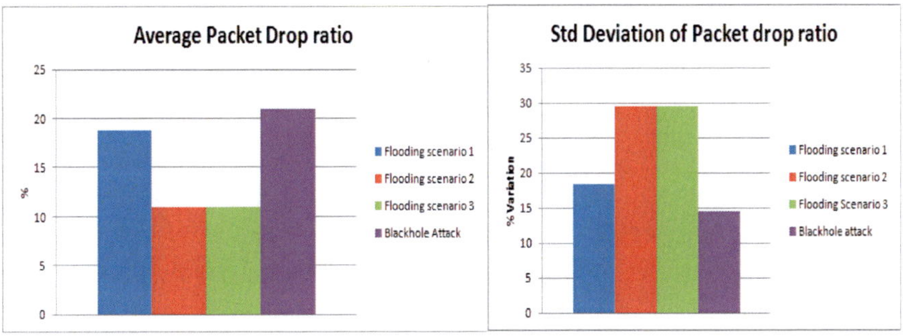

Figure 66: Comparison Packet Drop ratio (Flooding and Black Hole Attack)

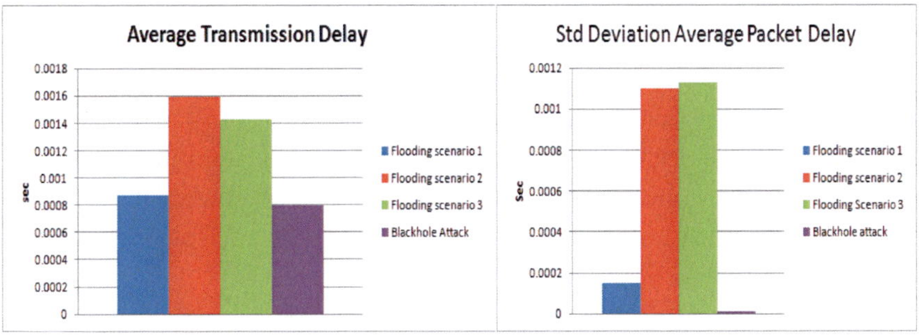

Figure 67: Comparison Average packet delay (Flooding and Black Hole Attack)

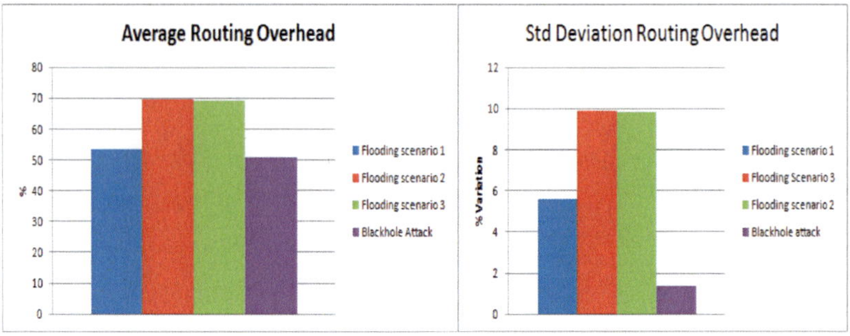

Figure 68: Comparison Routing Overhead (Flooding and Black Hole Attack)

Comparison Summary:

We plotted the standard deviation of all the three network performance parameters for all the attack scenarios (see Figures 68, 69, and 70). We have used averages of the results and standard deviation. This helps us to interpret the results decisively and come to a more definite conclusion.

Packet Drop Ratio:

The average of the packet drop ratio for Black Hole attack is slightly more than the three Flooding scenarios. The results are more comparable between the Flooding broadcast (19%) and the Black Hole attack (21%). On the other hand when we observed the standard deviation, the Black Hole attacks tend to vary less than the Flooding attacks. This shows that with increasing intensity of Flooding attack the packet drop ratio will increase.

Average Delay:

The average packet transmission delays in Flooding attacks are quite high in comparison to the Black Hole attacks. The Flooding transmission delay averages are in the range 0.8 μsec to 1.6 μsec, while that of Black Hole attack is 0.7 μsec. In the case of the standard deviation, the transmission delays in flooding attacks vary drastically with higher rates (0.2 μsec to 1 μsec of flooding packets) while Black Hole values are in the range of 10 μsec. This shows that the Flooding attack has more impact on the transmission delays of MANET network.

Routing Overhead:

The average routing overhead of the Black Hole attack is around 50%, while the Flooding attack average routing overhead range between 53% to 65% and this is more than that of Black Hole attack. The standard deviation of the routing averages is also much less for Black Hole attack (approx. 2%) compared to Flooding attacks (5% to 10%).
From all these three results, we concluded that Black Hole attack is less effective attack than Flooding attack. The obvious reason is that the Black Hole node only drops routing messages and do not take up the precious network bandwidth; while the Flooding not only sends malicious data but also wrong routing messages, and this consumes network bandwidth and processing power. This keeps the nodes busy in unnecessary processing and there is a long queue for valid data messages. This takes extra transmission times due to network choking and long back off times due to unsuccessful transmissions.

5.2 Prevention Schemes

From our findings in the last section, we have come to a conclusion that the Flooding attacks have a more severe impact on the performance of MANETs. Hence, we will now look into two popular Flooding prevention schemes. Then, we will implement these schemes using NS-3, and keeping the same three Flooding scenarios. After that, we will observe improvement in the network performance due to these prevention schemes. The two prevention techniques that we have chosen are FAP (***Flooding Attack Prevention***) and AMTT (***Avoiding Mistaken Transmission Table***).

5.2.1 FAP (Flooding Attack Prevention)

FAP is a popular scheme and it deals with both data and routing Flooding schemes. This scheme describes a classical scenario of a Flooding attack as referred to in the papers mentioned in the section 2.3.4. They propose a simple scheme to prevent Flooding attack. For the case of data packet flooding, the authors have proposed the use of application layer to identify the Internet layer of the Flooding node and then in turn the routing protocol, such as AODV, to declare the route to this node as not only malicious but also remove it from the routing tables. This is called path cut-off and in this implementation we have used this scheme.

To implement this scheme in NS-3, we used the AODV routing packets to extract the information at the destination node. Since each AODV node contains the header of the application layer as well, we extracted this information and used it to compare with the binded sockets on the destination node. Due to the nature of prevention, this scheme is not useful for implementation at any intermediate node but only at the destination node. A code snippet is given below to compare the malicious packet header and determine that the packet is not from a valid node.

The destination node will mark the packets with high frequency and compare its destination port number with any unreachable port number and then identify the other nodes using a broadcast RERR message.

```
/************** if the packet is for local delivery **********/
  // Broadcast local delivery/forwarding
  for (std::map<Ptr<Socket>, Ipv4InterfaceAddress>::const_iterator j =
       m_socketAddresses.begin (); j != m_socketAddresses.end (); ++j)
    {
     Ipv4InterfaceAddress iface = j->second;

     if (m_ipv4->GetInterfaceForAddress (iface.GetLocal ()) == iif)
       if (dst == iface.GetBroadcast () || dst.IsBroadcast ())
         {
          if (m_dpd.IsDuplicate (p, header))
            {

             NS_LOG_DEBUG ("Duplicated packet " << p->GetUid ()
<< " from " << origin << " DropCount "<< ++Dupl_pkt_count);
             ++Dupl_pkt_count;
             return true;
            }

          UpdateRouteLifeTime (origin, ActiveRouteTimeout);// check
for route life time

          if (lcb.IsNull () == false)
            {
             NS_LOG_LOGIC ("Broadcast local delivery to " <<
iface.GetLocal ());
```

```
            //Ipv4InterfaceAddress iface = m_ipv4-
>GetInterfaceForAddress (iface.GetLocal ());
             Ptr<Packet> packet = p->Copy ();// copy original packet to
another packet
            UdpHeader udpHeader;
            pack->RemoveHeader (udpHeader);// extract udp header
            Ptr<Socket> socket = FindSocketWithInterfaceAddress
(iface);// find the interface for the local binded socket
            Address to;
            Ptr<Node> node = GetObject<Node> () ;

            uint32_t a = socket->GetSockName(to);


            InetSocketAddress iaddr_rcv =
InetSocketAddress::ConvertFrom (to);
            iaddr_rcv = iaddr_rcv;
           // if(iaddr_rcv.GetPort()!= 654)
            if(point.LookupPortLocal(80))// if packet for local delivery
                                   {
            std::cout<<"Interface destination port "<< "
"<<iaddr_rcv.GetPort()<<"\n";
                                   }
                                   else
                                   {

           if(udpHeader.GetDestinationPort()==Dest_unrechable)// if
packet is unreachable
             {


                        RecvError (packet, "255.255.255.255" ); //
Broadcast a RERR message to all the nodes about this packet
                         NS_LOG_DEBUG ("Drop packet " << p->GetUid
() << " because no route to forward it.");
                         return false; // return from the route input with a
false state for treatment by the lower layers



             }
                                   }
```

```
            lcb (p, header, iif);// if the packet is valid fall through to
additional processing
          }
/************* end of FAP
***************************************/
```

5.2.1.1 Scenario1 (Flooding Broadcast -Message Packet Broadcast)

Table 5: Statistics of Network Performance after FAP Implementation for Scenario 1

Flooding Interval (sec)	No. of Sent Packets	No. of Received packets	Packet Drop Ratio (%)	Time Delay	Average Packet delay (sec)	Routing overhead (%)
1.2	137	1052	4.0145985	0.874698	0.000831462	65.72
1.1	137	1048	4.379562	0.869049	0.000829245	65.74
1	137	1047	4.4708029	0.874631	0.000835369	65.8
0.9	137	1042	4.9270073	0.872595	0.000837423	65.83
0.8	137	1022	6.7518248	0.853411	0.00083504	65.88
0.7	137	1020	6.9343066	0.8656	0.000848627	65.9
0.6	137	1010	7.8467153	0.872588	0.000863949	66.1
0.5	137	997	9.0328467	0.879457	0.000882103	66.23
0.4	137	967	11.770073	0.8635349	0.000893004	66.33
0.3	137	916	16.423358	0.846202	0.000923801	67.78
0.2	137	895	18.339416	0.883109	0.000986714	74
0.1	137	613	44.069343	0.90253	0.001472316	100

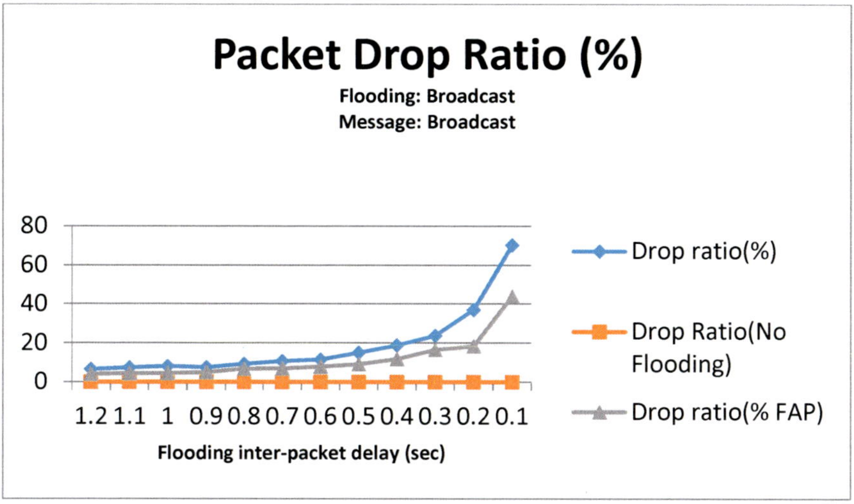

Figure 69: Packet Drop Ratio before and after FAP Implementation for Scenario1

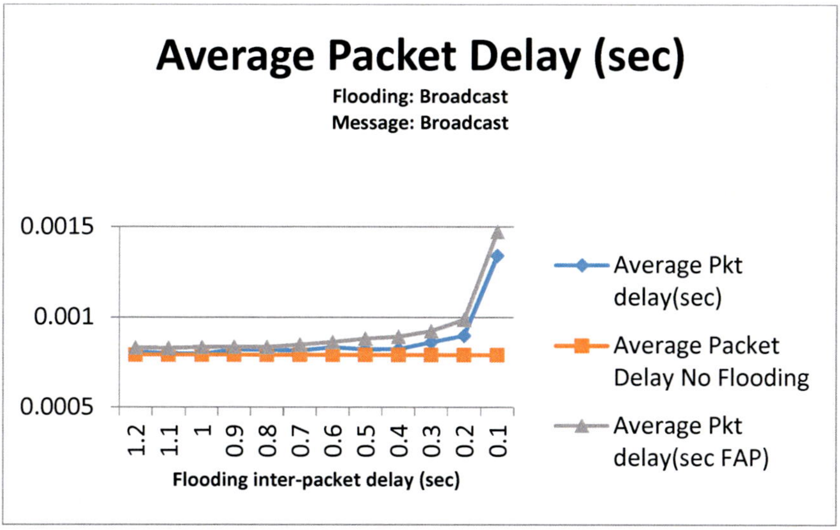

Figure 70: Average Packet Delay before and after FAP Implementation for Scenario1

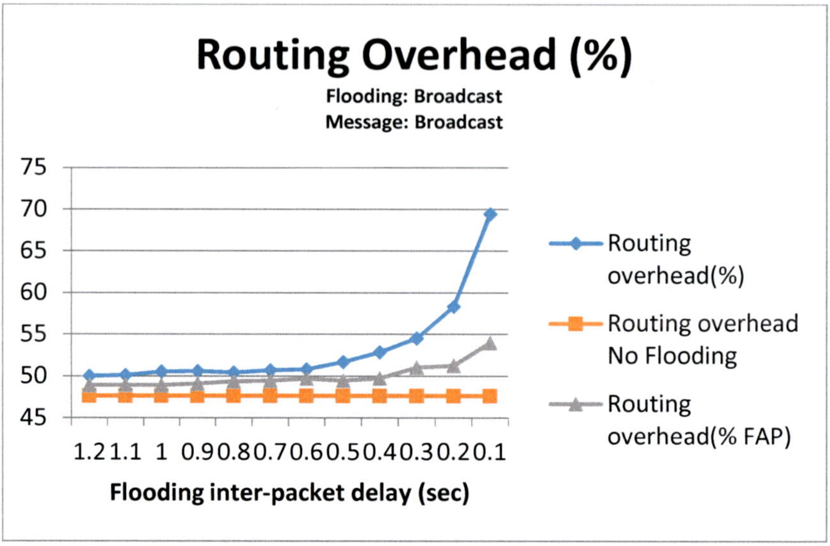

Figure 71: Routing Overhead before and after FAP Implementation for Scenario1

Summary of Scenario 1:

The above figures show that FAP can improve the number of dropped packets. In the worst case scenario of Flooding packet of 0.1 sec inter packet transmissions, the packet drop ratio was observed to be 70%, and this has improved to 40% which is quite effective. While the time delays have also seen improvements, now the packet average delays after implementing FAP are similar to the non-flooding scenarios, with low flooding volumes. Only in case of high Flooding attacks i.e. 0.1 and 0.2 seconds are the timing delays worse than when there was no FAP used. The obvious reason is that, due to high volumes of attack packets, the FAP is taking extra time to drop the unnecessary packets and send RERR AODV messages.

There is a significant improvement in the routing delays and this has improved to around 50% from the worst case scenario of 72% when FAP was not implemented on the Flooding scenario. Although RERR messages are being broadcast which increases the transmission delays, it reduces the routing overhead because the Flooding routes are deleted and hence unnecessary routing messages are curbed at neighboring nodes.

5.2.1.2 Scenario2 (Flooding Broadcast -Message Packet Unicast)

Table 6: Statistics of Network Performance after FAP Implementation for Scenario 2

Flooding Interval (sec)	No. of Sent Packets	No. of Received packets	Packet Drop Ratio (%)	Time Delay	Average Packet delay (sec)	Routing overhead (%)
1.2	137	137	0	0.135034	0.00098565	66.14
1.1	137	137	0	0.134755	0.000983613	65.74
1	137	137	0	0.137546	0.001003985	65.8
0.9	137	137	0	0.139246	0.001016394	65.83
0.8	137	137	0	0.138563	0.001011409	65.88
0.7	137	137	0	0.14972	0.001092847	65.9
0.6	137	137	0	0.143189	0.001045175	66.36
0.5	137	137	0	0.147377	0.001075745	66.23
0.4	137	137	0	0.150529	0.001098752	66.33
0.3	137	137	0	0.158859	0.001159555	67.78
0.2	137	97	29.19708029	0.123529	0.001273495	74.61
0.1	137	31	77.37226277	0.145732	0.005	82.34

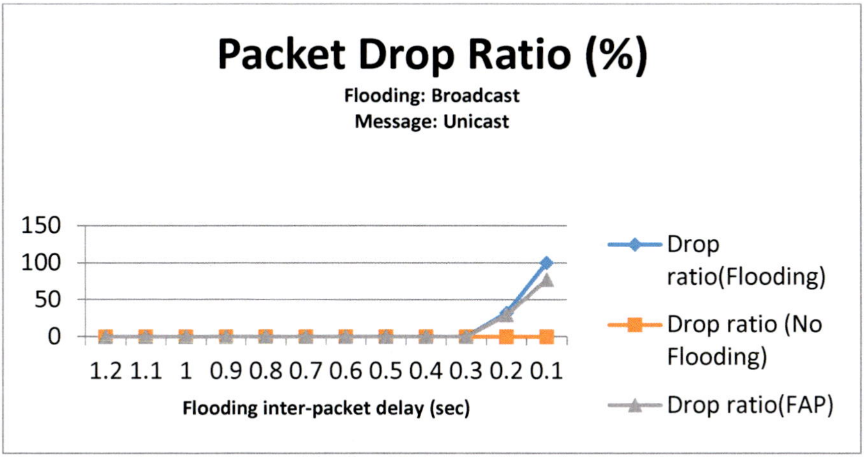

Figure 72: Packet Drop Ratio before and after FAP Implementation for Scenario2

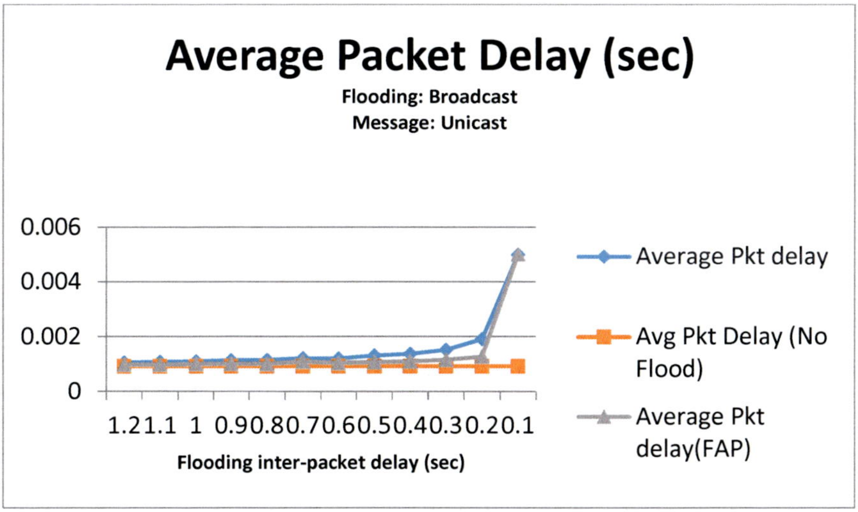

Figure 73: Average Packet Delay before and after FAP Implementation for Scenario2

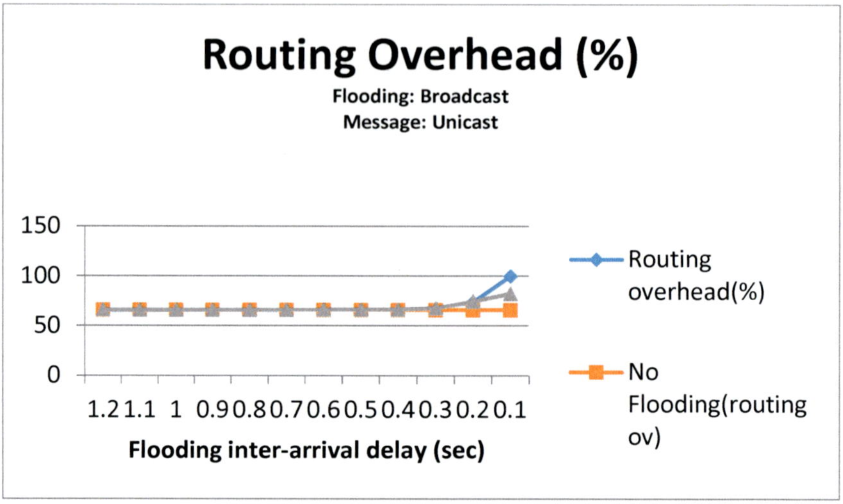

Figure 74: Routing Overhead before and after FAP Implementation for Scenario2

Summary of Scenario 2:

In the unicast transmission scenario, the FAP improves the packet drop ratio. In the worst case scenario of Flooding packet of 0.1 sec inter packet transmissions the packet drop ratio was observed to be 100%, this have been improved to 80% which is quite effective. While the time delays have also seen improvements, the packet average delays after implementing FAP are similar to the non-flooding scenarios, with low flooding volumes. In the case of high Flooding attacks i.e. 0.1 and 0.2 seconds the timing delays are better than when there was no FAP used.

There is a significant improvement in the routing delays and it has improved to around 82% from the worst case scenario of 100% when FAP was not implemented on the Flooding scenario. Although RERR messages are being broadcast which is increasing the transmission delays, it reduces the routing overhead because the Flooding routes are deleted and hence unnecessary routing messages are curbed at neighboring nodes.

5.2.1.3 Scenario3 (Flooding Unicast - Message Packet Unicast)

Table 7: Statistics of Network Performance after FAP Implementation for Scenario 3

Flooding Interval (sec)	No. of Sent Packets	No. of Received packets	Packet Drop Ratio (%)	Time Delay	Average Packet delay (sec)	Routing overhead (%)
1.2	137	137	0	0.136507	0.000996401	65.88
1.1	137	137	0	0.13954	0.00101854	65.9
1	137	137	0	0.14171	0.00103438	65.9
0.9	137	137	0	0.13964	0.00101927	65.9
0.8	137	137	0	0.1464	0.001068613	65.9
0.7	137	137	0	0.141733	0.001034547	65.9
0.6	137	137	0	0.144345	0.001053613	65.93
0.5	137	137	0	0.14738	0.001075766	65.92
0.4	137	137	0	0.158742	0.001158701	65.98
0.3	137	137	0	0.165669	0.001209263	65.86
0.2	137	101	26.27737226	0.13792	0.001365545	73.98
0.1	137	37	72.99270073	0.13891	0.003754324	77

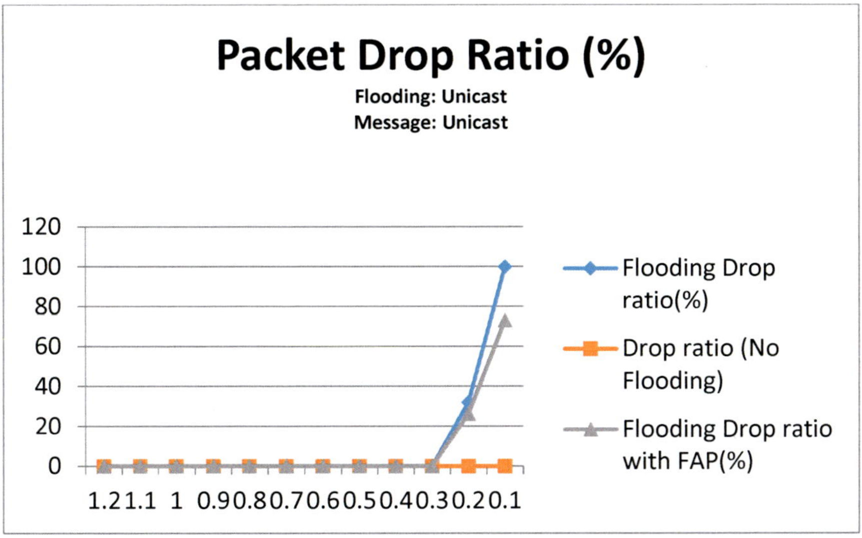

Figure 75: Packet Drop Ratio before and after FAP Implementation for Scenario3

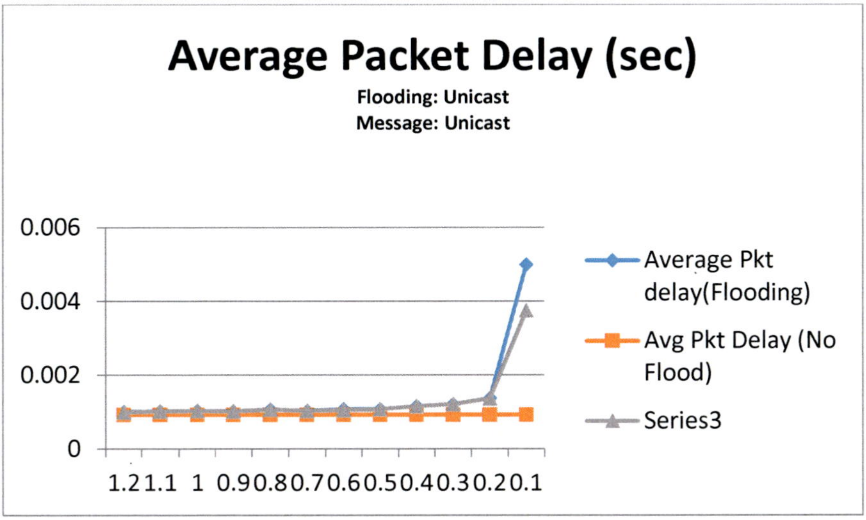

Figure 76: Average Packet Delay before and after FAP Implementation for Scenario3

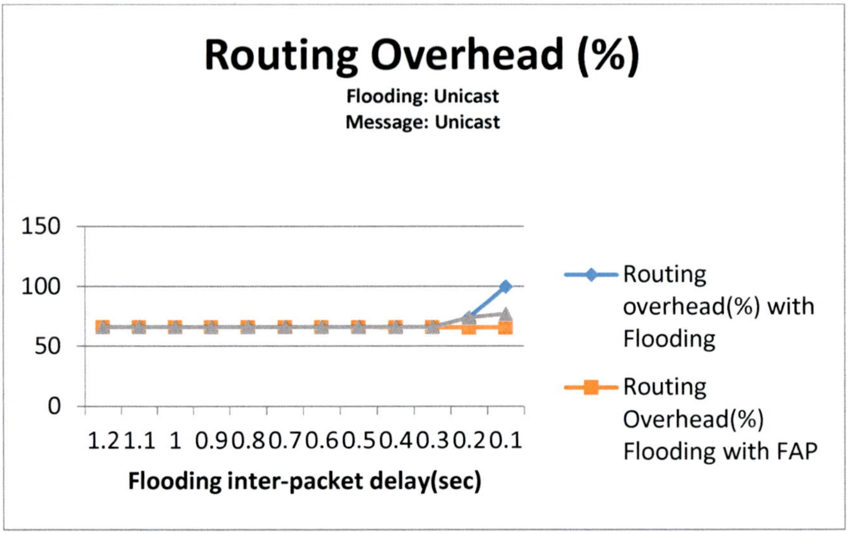

Figure 77: Routing Overhead before and after FAP Implementation for Scenario3

Summary of Scenario 3:

In the unicast transmission and flooding scenario the FAP improves the packet drop ratio. In the worst case scenario of Flooding packet of 0.1 sec inter packet transmissions the packet drop ratio was observed to be 100%, this has been improved to 74%, which is quite effective. While the time delays have also seen improvements; now the packet average delays after implementing FAP are similar to the non-flooding scenarios, with low flooding volumes. In case of high Flooding attacks i.e. 0.1 and 0.2 seconds, the timing delays improved to 0.35 from 0.5ms.

There is also a significant improvement in the routing delays with the implementation of FAP. The routing delays have improved to around 74% from the worst case scenario of 100% when FAP was not implemented in the Flooding scenario. Although RERR messages are being broadcast which increases the transmission delays, it reduces the routing overhead because the Flooding routes are deleted and hence unnecessary routing messages are curbed by neighboring nodes.

5.2.2 AMTT (Avoiding Mistaken Transmission Table)

AMTT uses a separate table then its routing table to keep track of messages traversing through the network. The implementation of AMTT is a complex as it needs to capture and store a separate table from the main Routing table. This was done by creating a new structure with all the required fields such as source and destination IP addresses Sequence number and Valid Flag.

When a packet is received through the route input the entry is created in the AMTT table by first checking if there is any existing entry not present. If an entry is present then the Valid Flag only is updated. If the destination node, after sensing that a packet from particular node is not valid, is sent back a reply message with the Valid Flag as zero, this indicates to all the intermediate nodes that the route is invalid and the routing entry for this route is discarded while the node with that particular IP is entered into the black list.

This is quite an affective scheme but it is complex and the memory overhead with each node needs to contain a largely redundant routing table to store the validity of the transmitting node.

Again this scheme was tested under Flooding attack with increasing frequency of flooding packets. The results are shown in the next section. The graphs are presented to compare all the 4 different possibilities to compare the outcomes. Network performance parameters are plotted for No-Flooding, Flooding, Flooding with FAP prevention and Flooding with AMTT implementation. We have given summary for each different scenario at the end of the graphs but have not explained each value of the graph. The detailed values for each Flooding interval are given in the table for comprehensive understanding.

5.2.2.1 Scenario1 (Flooding Broadcast -Message Packet Broadcast)

Table 8: Statistics of Scenario 1 with Flooding Attack and AMTT Prevention

Flooding Interval (sec)	No. of Sent Packets	No. of Received packets	Packet Drop Ratio (%)	Time Delay	Average Packet delay (sec)	Routing overhead (%)
1.2	137	1051	4.1058394	1.0533	0.001002188	49.13
1.1	137	1042	4.9270073	1.0732	0.001029942	49.23
1	137	1041	5.0182482	1.0345	0.000993756	50.13
0.9	137	1040	5.1094891	1.0453	0.001005096	50.45
0.8	137	1033	5.7481752	1.01	0.000977735	50.61
0.7	137	1031	5.9306569	1.02	0.000989331	51.02
0.6	137	1027	6.2956204	1.034	0.001006816	51.49
0.5	137	1010	7.8467153	1.01	0.001	51.6
0.4	137	985	10.127737	1.01	0.001025381	51.98
0.3	137	931	15.054745	1.01	0.001084855	52.09
0.2	137	909	17.062044	1.01	0.001111111	53.27
0.1	137	683	37.682482	1.01	0.00147877	55.63

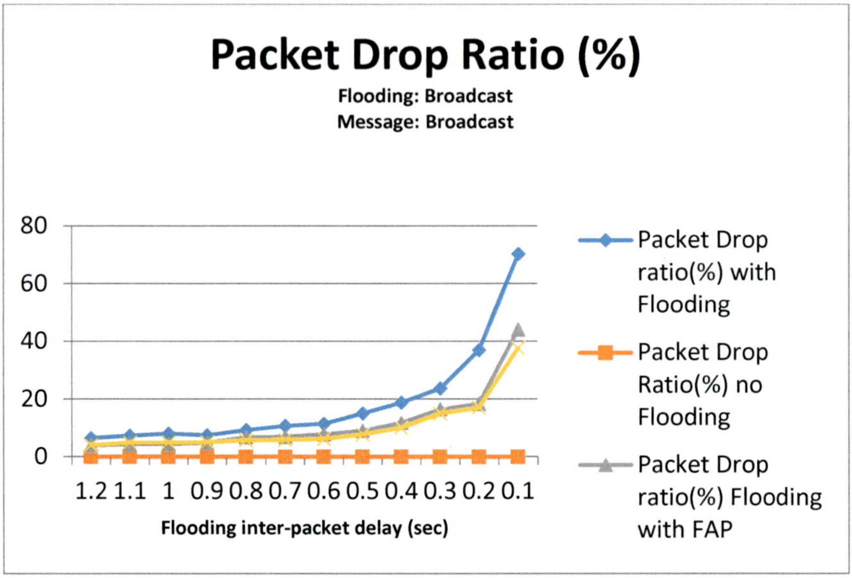

Figure 78: Comparison of Packet Drop Ratio Statistics

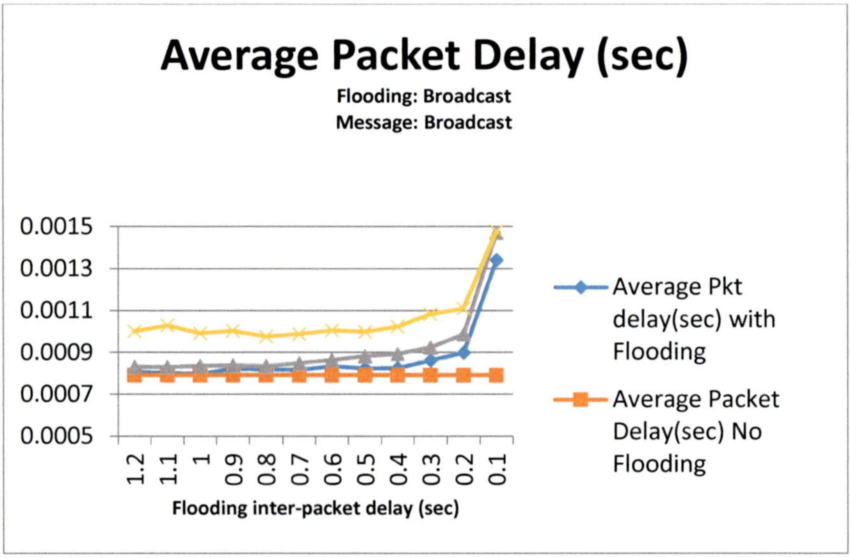

Figure 79: Comparison of Average Packet Delay

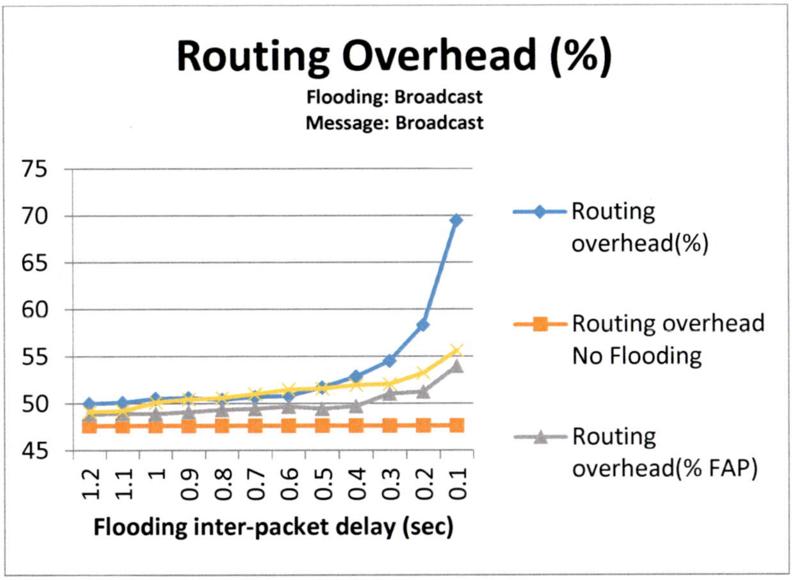

Figure 80: Comparison of Routing Overhead Statistics

Summary of Scenario 1:

- The packet drop ratio shows an improvement when used with AMTT as compared to FAP, this shows that the network have better throughput when AMTT is implemented. AMTT keeps a detailed track and ensures that the invalid routes are flagged and discarded even if new connections from the flooding nodes are created.

- The average timing delays suffer in the case of AMTT, since it takes longer because of extra processing at each node. Now not only each node compares the entries in the main routing tables but same process also take place for the AMTT; this increase the time delay by a significant margin. The memory usage must have also been increased but no concrete values are calculated as it is not in the scope of the current project.

- The Routing overhead also increased a little since extra route replies are needed. This is similar to the FAP implementation as not many extra routing messages are generated.

5.2.2.2 Scenario2 (Flooding Broadcast -Message Packet Unicast)

Table 9: Statistics of Scenario 2 with Flooding Attack and AMTT Prevention

Flooding Interval (sec)	No. of Sent Packets	No. of Received packets	Packet Drop Ratio (%)	Time Delay	Average Packet delay (sec)	Routing overhead (%)
1.2	137	137	0	0.135034	0.00098565	66.14
1.1	137	137	0	0.134755	0.000983613	65.74
1	137	137	0	0.137546	0.001003985	65.8
0.9	137	137	0	0.139246	0.001016394	65.83
0.8	137	137	0	0.138563	0.001011409	65.88
0.7	137	137	0	0.14972	0.001092847	65.9
0.6	137	137	0	0.143189	0.001045175	66.36
0.5	137	137	0	0.147377	0.001075745	66.23
0.4	137	137	0	0.150529	0.001098752	66.33
0.3	137	137	0	0.158859	0.001159555	67.78
0.2	137	97	29.19708029	0.123529	0.001273495	74.61
0.1	137	31	77.37226277	0.145732	0.005	82.34

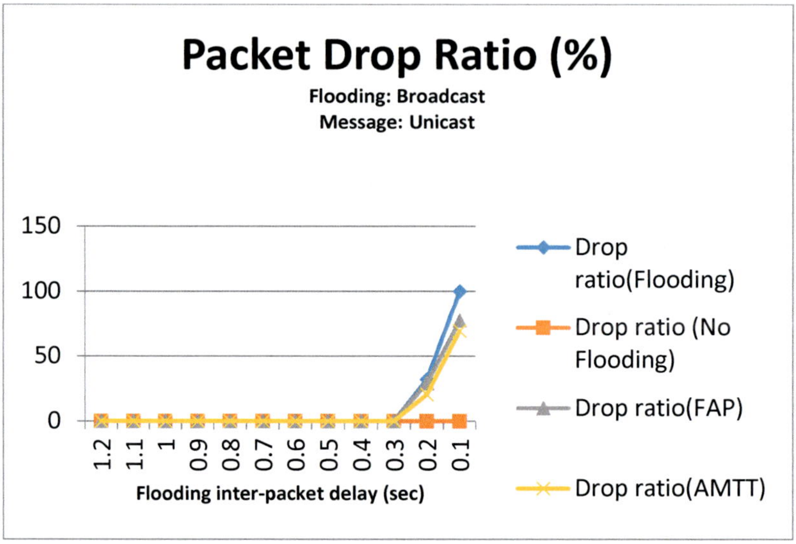

Figure 81: Comparison of Packet Drop Ratio for Scenario2

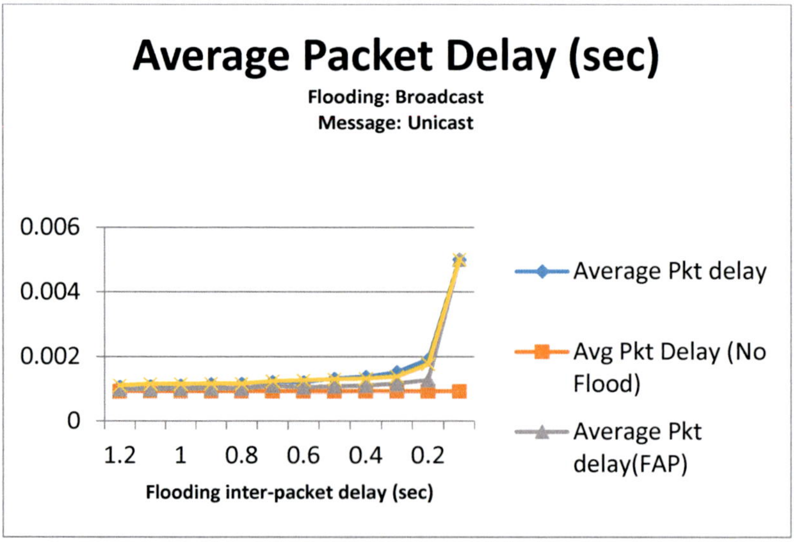

Figure 82: Comparison of Average Packet Delay in Scenario2

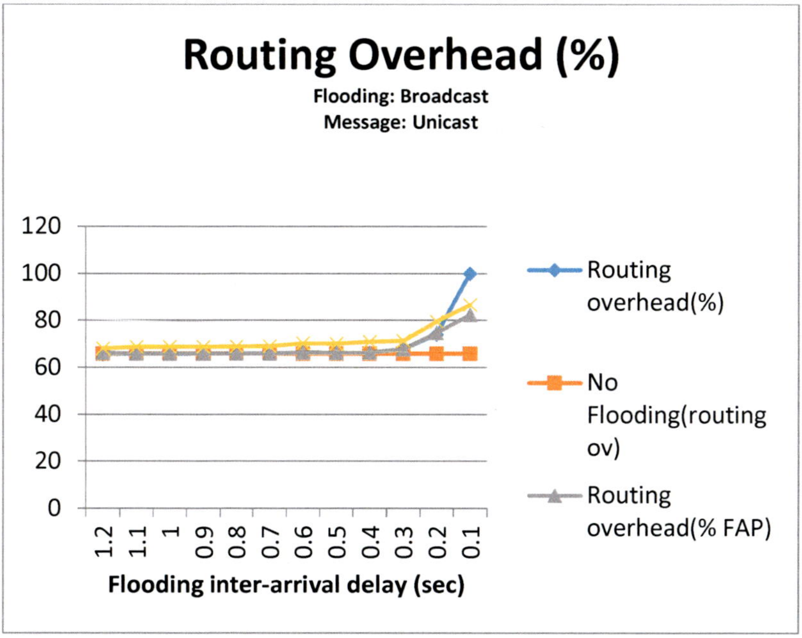

Figure 83: Comparison of Routing Overhead for Scenario2

Summary of Scenario 2:

- The packet drop ratio shows an improvement when used with AMTT compared to FAP, since for low frequency of flooding packets the drop ratio is almost zero. The only significant improvement is seen when the flooding packet frequency is 0.2 sec to 0.1 sec. The AMTT has shown an improvement of approximately 9% to 10%.

- The average packet delays have also increased in this scenario for AMTT compared to FAP. The reason is same as there is more processing time at each node due to keeping the AMTT.

- The Routing overhead also increased a little since extra route replies are needed. This is similar to the FAP implementation as not many extra routing messages are generated.

5.2.2.3 Scenario 3 (Flooding Unicast - Message Packet Unicast)

Table 10: Statistics of Scenario 3 with Flooding Attack and AMTT Prevention

Flooding Interval (sec)	No. of Sent Packets	No. of Received packets	Packet Drop Ratio (%)	Time Delay	Average Packet delay (sec)	Routing overhead (%)
1.2	137	137	0	0.135034	0.00098565	65.92
1.1	137	137	0	0.134755	0.000983613	66.01
1	137	137	0	0.137546	0.001003985	66.02
0.9	137	137	0	0.139246	0.001016394	66.01
0.8	137	137	0	0.138563	0.001011409	66
0.7	137	137	0	0.14972	0.001092847	66.23
0.6	137	137	0	0.143189	0.001045175	66.24
0.5	137	137	0	0.147377	0.001075745	66.71
0.4	137	137	0	0.150529	0.001098752	66.8
0.3	137	137	0	0.158859	0.001159555	68.1
0.2	137	107	21.89781022	0.123529	0.001154477	74.24
0.1	137	49	64.23357664	0.145732	0.002974122	77.32

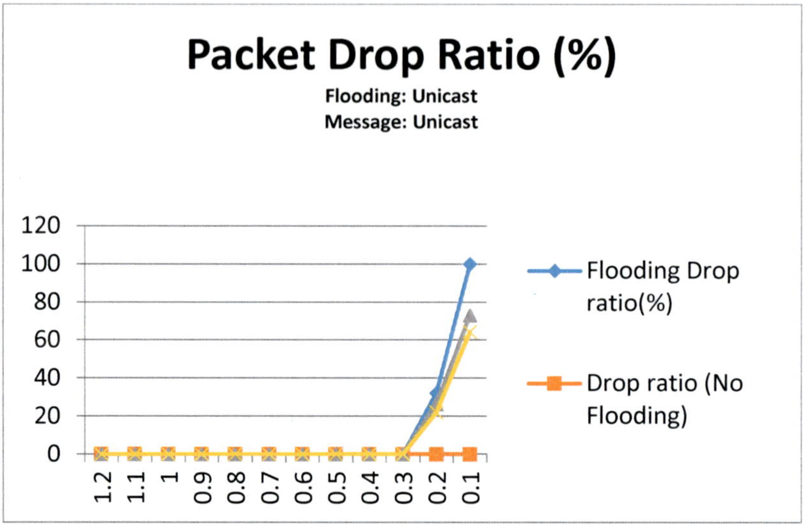

Figure 84: Compariosn of Packet Drop ratio for Flooding with Prevention of FAP or AMTT in Scenario3

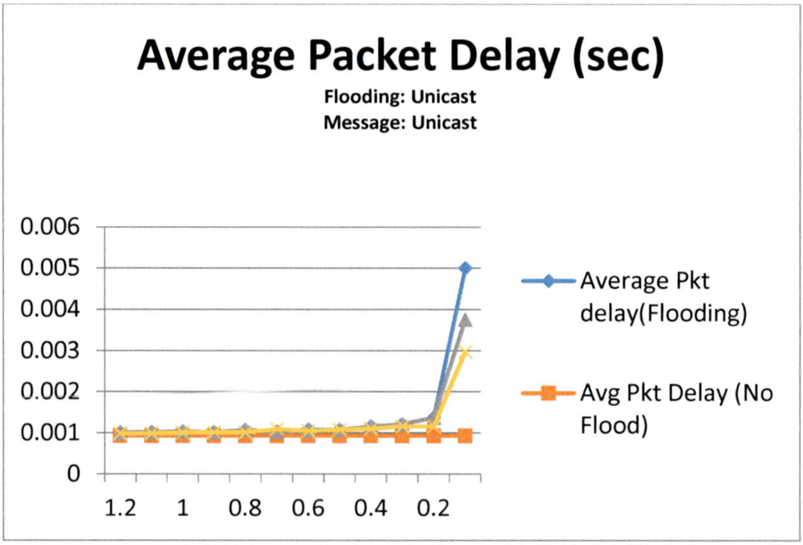

Figure 85: Average Packet Delay Comparison for Flooding with Prevention of FAP or AMTT in Scenario3

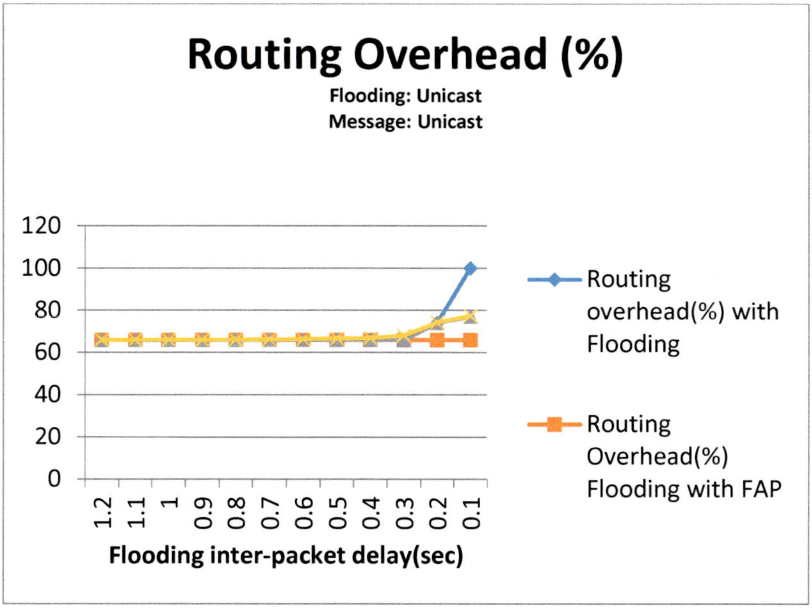

Figure 86: Routing Overhead for Flooding with Prevention of FAP or AMTT in Scenario3

Summary of Scenario 3:

- The packet drop ratio shows an improvement when used with AMTT compared to FAP, since for low frequency of flooding packets the drop ratio is almost zero. The only significant improvement is seen when the flooding packet frequency is 0.2 to 0.1 sec. The AMTT has shown an improvement of about 5% to 7%.

- The average packet delays have also increased in this scenario for AMTT compared to FAP. The reason is same as there is more processing time at each node due to keeping the AMTT. The average delays have increased by a value of 0.3 to 0.6ms.

- The Routing overhead also increased a little since extra route replies are needed. This is similar to the FAP implementation as not many extra routing messages are generated.

5.3 Conclusion and Future Work

5.3.1 Conclusion

This work deals with monitoring the AODV protocol performance for use in MANETs. We showed different scenarios of Flooding attack and Black Hole attack on a MANET with 10 nodes which are connected on wireless network with random walk.

The network performance was evaluated based on packet drop ratio, average packet delay and routing overhead with implementation of no attack and attack scenarios in ns-3. We chose three different Flooding attacks with broadcast and unicast packets. It was observed that the Flooding attack severely deteriorates the network performance. A packet drop ratio in the flooding scenarios was observed at between 70% and 100% in the unicast scenario. The Black Hole in comparison had a lower packet drop ratio. This was also observed for average packet delay, and routing overhead the Flooding attack had a severe effect on the network performance. It was decided to look into different prevention schemes which can be effective for countering Flooding attacks.

We then chose two existing techniques, FAP and AMTT. The results from the FAP showed up to 30 to 35% improvement in packet drop ratio. The average packet delay in some scenarios increases when FAP prevention is used because of the extra processing time to segregate Flooding packets from valid data packets. The routing overhead improves when FAP is employed because the scheme identifies the route from the Flooding nodes as malicious and the routing packets, then at the intermediate nodes, are dropped.

Similarly with the implementation of AMTT the packet drop ratio has shown some improvement in comparison with FAP. There is an average of 6 to 10% decrease in the packet drop ratio, but this improvement comes with a penalty. The timing delays have increased to a level of 0.5 to 0.7ms; also the complexity and the memory usage of this protocol is quite high. On the other hand the routing overhead has not increased to very significant levels when compared to the FAP implementation. This is because a similar amount of routing updates and error messages are required as in FAP, and no significant overhead is added to the protocol.

5.3.2 Future Work

From the results observed during Flooding attacks and their prevention schemes, it was observed that the packet drop ratio increases significantly with a high number of flooding packets. The main reason behind this is due to the flooding nodes gaining access to the transmission slots and trying to waste precious network bandwidth. Although the FAP and AMTT implementation shows significant improvement in the throughput of the network, they are still not good enough. So, in our observation, if these prevention schemes are used in conjunction with medium access schemes, this will greatly increase the performance. For example if a node is unable to transmit because of an existing transmission, and then it is found that the transmission was related to a malicious or flooded packet, the transmission back off times should not be as severe as for an actual scenario.

References

[1] M. Sreedevi and R. Seshadri, "An Innovative Kind Of Security Protocol Using Fusion Encryption In Virtual Private Networking," *International Journal*, vol. 3, no. 1, pp. 195–206, 2012.

[2] A. Ganz, Q. Xue, and H. Xu, "Adaptive Congestion Control in Infrastructure Wireless LANs with Bounded Medium Access Delay," *International Mobility and Wireless Access Workshop*, pp. 44–49.

[3] A. Tewari, U. Srivastava, and D. Sanghi, "A Distributed Monitoring System for troubleshooting Wireless Networks," no. Section 2, 2002.

[4] A. Srivastava, J. Singh, and S. Diwakar, "Wireless Lan and Ieee 802.11 in Digital Library," pp. 8–10, 2007.

[5] R. McConnell, "Testing and Evaluation of Shipboard Wireless Network Components," 2000.

[6] I. Chlamtac, M. Conti, and J. J.-N. Liu, "Mobile ad hoc networking: imperatives and challenges," *Ad Hoc Networks*, vol. 1, no. 1, pp. 13–64, Jul. 2003.

[7] R. Ramanathan and J. Redi, "A brief overview of ad hoc networks: challenges and directions," *IEEE communications Magazine*, no. May, pp. 20–22, 2002.

[8] M. Murty and M. Das, "Performance Evalution of MANET Routing Protocols using Reference Point Group Mobility and Random WayPoint Models," *International Journal of Ad hoc, Sensor & Ubiquitous ...*, vol. 2, no. 1, pp. 33–43, 2011.

[9] A. Bandyopadhyay, "A simulation analysis of flooding attack in MANET using NS-3," ... *Information Theory and ...*, pp. 1–5, Feb. 2011.

[10] S. Upadhyay, P. Joshi, N. Gandotra, and A. Kumari, "Comparison and performance analysis of reactive type DSR, AODV and proactive type DSDV routing protocol for wireless mobile ad-hoc network, using NS-2 simulator," *Journal of Engineering and Computer Innovations*, vol. 3, no. 2, pp. 36–47, Mar. 2012.

[11] Y. Guo, S. Gordon, and S. Perreau, "A Flow Based Detection Mechanism against Flooding Attacks in Mobile Ad Hoc Networks," *IEEE Wireless Communications and Networking Conference*, pp. 3105–3110, 2007.

[12] H. Aldabbas and T. Alwada'n, "Data Confidentiality in Mobile Ad hoc Networks," *Arxiv preprint arXiv: ...*, vol. 4, no. 1, pp. 225–236, 2012.

[13] M. Scholar, "Multicast Multi-Path Power Efficient Routing In Mobile Adhoc Networks," *ijmttjournal.org*, vol. 3, pp. 147–149, 2012.

[14] C. S. Sahin, "Bio-Inspired Techniques for Topology Control of Mobile Nodes," pp. 217–219.

[15] S. Basagni, M. Conti, and S. Giordano, *Mobile ad hoc networking*. 2004, pp. 1–63.

[16] S. Hussain, "Dynamic Route Shortening and Route Repairing Mechanism for Mobile Ad Hoc Networks," *Journal of Computer Science*, vol. 8, no. 8, pp. 1212–1218, 2012.

[17] S. Revathi and T. Rangaswamy, "Secured Optimal Adaptable Ad-Hoc Routing Protocol in MANET," *European ...*, vol. 73, no. 1, pp. 62–72, 2012.

[18] Q. Zeng and D. Agrawal, "Handbook of wireless networks and mobile computing," *Chapter*, vol. 8, pp. 0–471, 2002.

[19] P. Malik, "Consequences of limited manet resources," *International Journal*, vol. 2, no. 1, pp. 2–4, 2012.

[20] J. Wu and I. Stojmenovic, "Ad hoc networks," *COMPUTER-IEEE COMPUTER SOCIETY-*, no. February, pp. 29–31, 2004.

[21] M. B. Yassein and L. Abu-Fadaleh, "New Probabilistic-Based Broadcasting Algorithm for Mobile Ad Hoc Networks," *ICIMP 2011, The Sixth International ...*, no. c, pp. 45–49, 2011.

[22] G. Varaprasad, S. Dhanalakshmi, and M. Rajaram, "New security algorithm for mobile adhoc networks using zonal routing protocol," *Ubiquitous Computing and ...*, 2008.

[23] A. Patil, N. Sambaturu, and K. Chunhaviriyakul, "Convergence Time Evaluation of Algorithms in MANETs," *Arxiv preprint arXiv:0910.1475*, vol. 5, no. 1, pp. 144–149, 2009.

[24] V. Kärpijoki, "Security in ad hoc networks," *Helsinki University of Technology*, pp. 1–8, 2000.

[25] P. Anupriya and J. Suba, "ANT – AODV Hybrid Protocol for MANETs," vol. 1, no. 1, pp. 3–7, 2012.

[26] K. Biswas and M. Ali, "Security threats in mobile ad hoc network," *University essay from Blekinge Tek nisha Ho ...*, vol. 1, no. 2, pp. 125–129, 2007.

[27] M. Kefayati, "Key Management in Mobile Ad Hoc Networks," 2005.

[28] Z. Luying, B. Jit, and T. Enghwa, "A Bridging Method for Mobile Ad hoc Networks," *ukpmc.ac.uk.*

[29] Shaveta, "Pursue Shortest Mobility Model with Delay and Retransmission Attempts in MANET," *International Journal of Electronics and Computer Science Engineering*, vol. 1, no. 3, pp. 1167–1175, 2012.

[30] R. Nichols and P. Lekkas, *Wireless security*. 2002.

[31] J. Jamal, "MANET (Mobile ad hoc network) - Characteristics and Features." [Online]. Available: http://www.eexploria.com/manet-mobile-ad-hoc-network-characteristics-and-features/.

[32] J. Haillot and F. Guidec, "Towards a Usenet-Like Discussion System for Users of Disconnected MANETs," *22nd International Conference on Advanced Information Networking and Applications - Workshops (aina workshops 2008)*, pp. 1678–1683, 2008.

[33] P. Papadimitratos and Z. J. Haas, "Secure message transmission in mobile ad hoc networks," *Ad Hoc Networks*, vol. 1, no. 1, pp. 193–209, Jul. 2003.

[34] S. Velagaleti, M. Seetha, and S. Raju, "Design of a Scheme for Secure Rouitng in Mobile Adhoc Networks," *ijcttjournal.org*, pp. 251–254, 2011.

[35] I. Ullah and S. Rehman, "Analysis of Black Hole attack on MANETs Using different MANET routing protocols," *Program Electrical Engineering with emphasis on ...*, 2010.

[36] A. Nayyar, "Detecting Sequence Number Collector Problem in Black Hole Attacks in AODV Based Mobile Adhoc Networks," *International Journal of Advanced Research in ...*, vol. 1, no. 4, pp. 525–529, 2012.

[37] A. A.Vani and D. S. Rao, "Providing of Secure Routing against Attacks in MANETs," *International Journal of Computer Applications*, vol. 24, no. 8, pp. 16–25, Jun. 2011.

[38] W. Bhaya and S. Alasadi, "Security against Spoofing Attack in Mobile Ad Hoc Networks," *European Journal of Scientific Research*, vol. 64, no. 4, pp. 634–643, 2011.

[39] V. Devi and R. Bhuvaneswaran, "Adaptive Association Rule Mining Based cross Layer Intrusion Detection System for MANET," *International Journal*, vol. 3, no. 5, 2011.

[40] H. L. Nguyen and U. T. Nguyen, "Study of Different Types of Attacks on Multicast in Mobile Ad Hoc Networks," *International Conference on Networking, International Conference on Systems and International Conference on Mobile Communications and Learning Technologies (ICNICONSMCL'06)*, pp. 149–149, 2006.

[41] S. Vincent, "Improving CA-AOMDV Protocol Against Blackhole Attacks," *jcaksrce.org*, 2012.

[42] D. Roy and R. Chaki, "MADSN: Mobile Agent Based Detection of Selfish Node in MANET," *... Journal of Wireless & Mobile Networks (IJWMN) Vol*, vol. 3, no. 4, pp. 225–235, 2011.

[43] V. Tafti and A. Gandomi, "Performance of QoS Parameters in MANET Application Traffics in Large Scale Scenarios," *World Academy of Science, Engineering and ...*, pp. 855–858, 2011.

[44] Z. Zhang and H. Zhou, "Empirical examination of Mobile Ad Hoc Routing Protocols on wireless sensor networks," *International Journal of Computer Networks ...*, vol. 1, no. 1, pp. 75–87, 2009.

[45] S. Ade and P. Tijare, "Performance Comparison of AODV, DSDV, OLSR and DSR Routing Protocols in Mobile Ad Hoc Networks," *International Journal of Information Technology ...*, vol. 2, no. 2, pp. 545–548, 2010.

[46] Y. Kim, S. Ahn, Y. Lee, J. Jeong, and J. Lee, "DESIGN AND IMPLEMENTATION OF IPV6 ADDRESS AUTOCONFIGURATION FOR AODV IN MOBILE AD HOC NETWORKS," 2008.

[47] E. Kulla, M. Ikeda, L. Barolli, R. Miho, and V. Kolici, "Effects of Source and Destination Movement on MANET Performance Considering OLSR and AODV Protocols," *2010 13th International Conference on Network-Based Information Systems*, pp. 510–515, Sep. 2010.

[48] A. Zahary and A. Ayesh, "On-demand multiple route maintenance in AODV extensions (ORMAD)," *2008 International Conference on Computer Engineering & Systems*, pp. 225–230, Nov. 2008.

[49] V. Rishiwal and S. Verma, "Energy Aware On Demand Routing for MANETs," *... and Sensor Networks, 2007. WCSN'07. ...*, pp. 89–94, 2007.

[50] R. Kumar, S. Kumar, S. Pradhan, and V. Yadav, "Modified route-maintenance in AODV Routing protocol using static nodes in realistic mobility model," *International Journal on …*, vol. 3, no. 4, pp. 1554–1562, 2011.

[51] H. Asenov and V. Hnatyshin, "GPS-Enhanced AODV routing," *Proc. 2009 International Conference on …*, 2009.

[52] H. A. Esmaili, "Performance Analysis of AODV under Black Hole Attack through Use of OPNET Simulator," vol. 1, no. 2, pp. 49–52, 2011.

[53] L. Tamilselvan and V. Sankaranarayanan, "Prevention of Co-operative Black Hole Attack in MANET," *Journal of Networks*, vol. 3, no. 5, pp. 13–20, May 2008.

[54] M. Alinci, E. Zanaj, and E. Kotobelli, "Effect of Black hole Attack on Mobile Ad-hoc Network Performance," pp. 1–6, 2012.

[55] T. P. Banga, V K and Singh, "Survey on Prevention of Black Hole Nodes in Mobile Adhoc Networks," pp. 105–109, 2012.

[56] S. CHAMOLI, S. KUMAR, and D. RANA, "Performance of AODV against Black Hole Attacks in Mobile ad-hoc Networks," *International Journal*, vol. 3, no. August, pp. 1395–1399, 2012.

[57] U. Khartad and R. Krishna, "Route Request Flooding Attack Using Trust based Security Scheme in Manet," *interscience.in*, no. 2248, pp. 27–33, 2012.

[58] M. PRADIP, M. MANGESH, and P. D. M. ALI, "A Survey of Mobile Ad Hoc Network Attacks,"

International Journal of ..., vol. 2, no. 9, pp. 4063–4071, 2010.

[59] S. Asherson and A. Hutchison, "Secure Routing for Wireless Mesh Networks," ... *Networks and Applications Conference ...*, 2006.

[60] K. Argyraki and D. Cheriton, "Scalable network-layer defense against internet bandwidth-flooding attacks," *IEEE/ACM Transactions on Networking (TON)*, pp. 1–14, 2009.

[61] M. HARI, "A NOVEL ROUTING ATTACK IN MOBILE AD HOC NETWORKS," *Indian Journal of ...*, vol. 1, no. 4, pp. 382–391, 2010.

[62] A. Al-Bayatti and H. Zedan, "Security Management for Mobile Ad Hoc Network of Networks (MANoN)," ... *Journal of Mobile ...*, no. February, 2010.

[63] W. Scott, A. Houle, and A. Martin, "Information assurance issues for an SDR operating in a manet network," *SDR Forum, November*, 2006.

[64] Y. Huang and W. Lee, "A cooperative intrusion detection system for ad hoc networks," *Proceedings of the 1st ACM workshop on Security of ad hoc and sensor networks - SASN '03*, p. 135, 2003.

[65] J. Sen, S. Koilakonda, and A. Ukil, "A Mechanism for Detection of Cooperative Black Hole Attack in Mobile Ad Hoc Networks," *2011 Second International Conference on Intelligent Systems, Modelling and Simulation*, pp. 338–343, Jan. 2011.

[66] K. Rana, "Methodology for Detecting and Thwarting DoS in MANET," *research.ijcaonline.org*, pp. 31–34, 2011.

[67] R. Karpaga Brinda and P. Chandrasekar, "Detection and Removal of Co-Operative Black Hole\Black Hole Attack in Manet," *International Journal of Computer Applications*, vol. 43, no. 11, pp. 5–9, Apr. 2012.

[68] M. Conti and S. Giordano, "Multihop ad hoc networking: The theory," *Communications Magazine, IEEE*, no. April, pp. 78–86, 2007.

[69] Y. Hu, S. Das, and H. Pucha, "Peer-to-peer overlay abstractions in manets," *... Aspects of Sensor, Ad Hoc Wireless and Peer-to-Peer ...*, pp. 857–874, 2005.

[70] C. T. Calafate, J. Cano, P. Manzoni, and M. P. Malumbres, "A QoS architecture for MANETs supporting real-time peer-to-peer multimedia applications," *Seventh IEEE International Symposium on Multimedia (ISM'05)*, pp. 193–200.

[71] C. Cheng and C. Wang, "A scalable location-aware peer-to-peer system for MANET," *... and Systems, 2005 2nd ...*, no. 5, 2005.

[72] P. Goyal, S. Batra, and A. Singh, "A Literature Review of Security Attack in Mobile Ad-hoc Networks," *International Journal of Computer ...*, vol. 9, no. 12, pp. 11–15, 2010.

[73] Z. Karakehayov, "Using REWARD to detect team black-hole attacks in wireless sensor networks," *Wksp. Real-World Wireless Sensor Networks*, 2005.

[74] S. Kurosawa and H. Nakayama, "Detecting blackhole attack on aodv-based mobile ad hoc networks by

dynamic learning method," *International Journal of ...*, vol. 5, no. 3, pp. 338–346, 2007.

[75] N. Bhalaji and a. Shanmugam, "Defense Strategy Using Trust Based Model to Mitigate Active Attacks in DSR Based MANET," *Journal of Advances in Information Technology*, vol. 2, no. 2, pp. 92–98, May 2011.

[76] A. Sharma and M. Sinha, "A Differential Evaluation Algorithm for routing Optimization in Mobile Ad-hoc Networks," vol. 1, no. 4, 2012.

[77] M. Ngadi, R. Khokhar, and S. Mandala, "A review current routing attacks in mobile ad-hoc networks," *International Journal of ...*, no. 2, pp. 18–29, 2008.

[78] H. Yang, H. Luo, F. Ye, S. Lu, and L. Zhang, "Security in mobile ad hoc networks: challenges and solutions," *IEEE Wireless ...*, no. February, pp. 38–47, 2004.

[79] P. Goyal, V. Parmar, and R. Rishi, "MANET: Vulnerabilities, Challenges, Attacks, Application," *International Journal of ...*, vol. 11, no. January, pp. 32–37, 2011.

[80] G. Padmavathi, P. Subashini, and M. Aruna, "ZRP with WTLS Key Management Technique to Secure Transport and Network Layers in Mobile Adhoc Networks," *International Journal*, vol. 4, no. 1, pp. 129–138, 2012.

[81] A. Kumar, "Security Attacks in Manet - A Review," 2011.

[82] A. Rai, R. Tewari, and S. Upadhyay, "Different Types of Attacks on Integrated MANET-Internet

Communication," *International Journal of ...*, no. 4, pp. 265–274, 2010.

[83] M. Karimi, "Quality of Service (QoS) Provisioning in Mobile Ad-Hoc Networks (MANETs)," in *Mobile Ad-Hoc Networks: Protocol Design*, .

[84] I. De and D. Roy, "Comparative study of Attacks on AODV-based Mobile Ad Hoc Networks," *International Journal*, vol. 3, no. 1, pp. 313–322, 2011.

[85] G. Padmaja and C. Lakshmi, "INTRUSION-TOLERANT ARCHITECTURE FOR MOBILE ADHOC NETWORKS," *ijeit.com*, vol. 1, no. 4, pp. 45–49, 2012.

[86] W. Ye, J. Heidemann, and D. Estrin, "Medium access control with coordinated adaptive sleeping for wireless sensor networks," *Networking, IEEE/ACM ...*, vol. 12, no. 3, pp. 493–506, 2004.

[87] T. Anantvalee and J. Wu, "A survey on intrusion detection in mobile ad hoc networks," *Wireless Network Security*, pp. 170–196, 2007.

[88] A. Bhosle, T. Thosar, and S. Mehatre, "Black-Hole and Wormhole Attack in Routing Protocol AODV in MANET," *Proceedings of IJCSEA*, vol. 2, no. 1, pp. 45–54, 2012.

[89] H. Pratihari, "Intrusion Detection System (IDS) for Secure MANETs: A Study," *ijera.com*, vol. 2, no. 1, pp. 962–966, 2012.

[90] H. Bathla and K. Lakhani, "A Novel Method for Intrusion Detection System to Enhance Security in Ad

hoc Network," *Arxiv preprint arXiv:1005.4023*, vol. 2, no. 5, pp. 101–107, 2010.

[91] L. KAPANG, P. Singh, and D. Sutradhar, "Comparative Study of Adhoc Routing Protocol AODV, DSR and DSDV in Mobile Adhoc NETwork," *Indian Journal of Computer Science and Engineering*, vol. 1, no. 4, pp. 364–371, 2011.

[92] J. Li and P. Mohapatra, *Quality of service (qos) provisioning in multihop ad hoc networks*, no. March. 2006.

[93] I. Teknologi Bandung, "MANET Deployment in Disaster Area." [Online]. Available: http://www.lppm.itb.ac.id/research/?p=411.

[94] N. O. Bouvin, "Peer-to-Peer Networking (P2P)."

[95] J. Hoebeke and I. Moerman, "An overview of mobile ad hoc networks: applications and challenges," *JOURNAL- ...*, pp. 60–66, 2004.

[96] "No Title," *Bluetronix*. [Online]. Available: http://www.bluetronix.net/products.htm.

[97] C. A. Ciufo, "A Mobile Starbucks Hotspot, in Spots That Aren't So Hot. And There's No Coffee." [Online]. Available: http://eecatalog.com/military/2012/07/30/a-mobile-starbucks-hotspot-in-spots-that-arent-so-hot-and-theres-also-no-coffee/.

[98] "No Title." [Online]. Available: http://www.grin.com/object/external_document.264643/ba8f2d99782d666f14dc4b63d90a19a8_LARGE.png.

[99] C.-K. Toh, "Maximum battery life routing to support ubiquitous mobile computing in wireless ad hoc networks," *IEEE Communications Magazine*, vol. 39, no. 6, pp. 138–147, Jun. 2001.

[100] R. Chadha and L. Kant, *Policy-driven mobile ad hoc network management*. 2007, pp. 1–24.

[101] D. Djenouri, L. Khelladi, and N. Badache, "A survey of security issues in mobile ad hoc networks," *IEEE communications surveys*, pp. 2–28, 2005.

[102] B. Wu, J. Chen, J. Wu, and M. Cardei, "A survey of attacks and countermeasures in mobile ad hoc networks," *Wireless Network Security*, pp. 1–38, 2007.

[103] S. Chakrabarti and A. Mishra, "Quality of service in mobile ad hoc networks," *The handbook of ad hoc wireless networks*, no. March, 2003.

[104] A. Shrivastava, A. Shanmogavel, and A. Mistry, "Overview of Routing Protocols in MANET's and Enhancements in Reactive Protocols," *cs.lamar.edu*.

[105] J. Eriksson, M. Faloutsos, and S. Krishnamurthy, "Routing Scalability in MANETs," *University of California, ...*, pp. 1–29, 2008.

[106] J. Yu and P. Chong, "A survey of clustering schemes for mobile ad hoc networks," *Communications Surveys & Tutorials, IEEE*, vol. 20, no. 20, pp. 135–151, 2005.

[107] D. Maltz, "On-demand routing in multi-hop wireless mobile ad hoc networks," pp. 19–36, 2001.

[108] M. Karimi and D. Pan, "Challenges for Quality of Service (QoS) in Mobile Ad-hoc Networks

(MANETs)," ... *Conference, 2009. WAMICON'09. IEEE 10th ...*, 2009.

[109] R. Asokan, "A review of Quality of Service (QoS) routing protocols for mobile Ad hoc networks," *Wireless Communication and Sensor Computing, ...*, 2010.

[110] C. Cordeiro and D. Agrawal, *Ad hoc and sensor networks: theory and applications*. 2011, pp. 1–17.

[111] F. Perich, a. Joshi, J. Undercoffer, T. Finin, L. Kagal, and Y. Yesha, "In reputation we believe: query processing in mobile ad-hoc networks," *The First Annual International Conference on Mobile and Ubiquitous Systems: Networking and Services, 2004. MOBIQUITOUS 2004.*, pp. 326–334.

[112] J. a. Stine and G. de Veciana, "A comprehensive energy conservation solution for mobile ad hoc networks," *2002 IEEE International Conference on Communications. Conference Proceedings. ICC 2002 (Cat. No.02CH37333)*, vol. 5, pp. 3341–3345.

[113] C. Srisathapornphat and C. Shen, "Coordinated Energy Conservation for Ad hoc Networks," *Communications*, 2002.

[114] M. Li, "Strategic Pricing to Stimulate Node Cooperation in Wireless Ad Hoc Networks Strategic Pricing to Stimulate Node Cooperation in Wireless Ad Hoc Networks," 2007.

[115] J. N. Al-Karaki and A. E. Kamal, "Stimulating Node Cooperation in Mobile Ad hoc Networks," *Wireless Personal Communications*, vol. 44, no. 2, pp. 219–239, Dec. 2007.

[116] D. Prince, A. Scott, and D. Shepherd, “Toward an efficient solution for dynamic ad hoc network interoperability,” *PGNet 2002*, 2002.

[117] “Protection of humans and the environment.” [Online]. Available: http://www.microdrones.com/company/research-and-development/uav-projects-microdrones.php.

[118] N. Meghanathan, “Survey and taxonomy of Unicast routing protocols for mobile ad hoc networks,” *University of Texas at Dallas, Richardson, TX, ...*, vol. 1, no. 1, pp. 1–21, 2004.

[119] H. Badis, A. Munaretto, K. A. Agha, and G. Pujolle, “QoS for Ad hoc Networking Based on Multiple Metrics: Bandwidth and Delay,” *Mobile and Wireless Communications Networks - Proceedings of the Fifth IFIP-TC6 International Conference*, pp. 15–18, 2003.

[120] “WolfTech Information Technology.” [Online]. Available: http://www.wolftech.ncsu.edu/towncrier/tcwriter/images/2012/05/fast-moving-mobile-networks1.jpg.

[121] M. Gerla, “Scalable routing protocols for mobile ad hoc networks,” *IEEE Network*, vol. 16, no. 4, pp. 11–21, Jul. 2002.

[122] K. Holter, “Comparing AODV and OLSR,” *folk. uio. no/kenneho/studies/essay/essay. html*, no. April, pp. 1–19, 2005.

[123] C. Liu and J. Kaiser, *A survey of mobile ad hoc network routing protocols*, no. 2003. 2003.

[124] Uma, "A comparative study and performance evaluation of reactive quality of service routing protocols in mobile ad hoc networks," 2009.

[125] C. Dhote and M. Pund, "Hybrid Routing Protocol with Broadcast Reply for Mobile Ad hoc Network," *International Journal of …*, vol. 1, no. 10, 2010.

[126] S. Saxena, B. Saran, and M. Singh, "Simulation And Performance Analysis of DSDV, OLSR, DSR and ZRP Routing Protocols in MANETs," *ijcttjournal.org*, vol. 3, no. 2, pp. 315–320, 2012.

[127] H. Amri, M. Abolhasan, and T. Wysocki, "Scalability of MANET routing protocols for heterogeneous and homogenous networks," *Computers & Electrical Engineering*, 2010.

[128] M. Abolhasan, T. Wysocki, and E. Dutkiewicz, "A review of routing protocols for mobile ad hoc networks," *Ad Hoc Networks*, vol. 2, no. 1, pp. 1–22, Jan. 2004.

[129] Y. H. D. Johnson and D. Maltz, "The Dynamic Source Routing Protocol (DSR) for Mobile Ad Hoc Networks for IPv4." [Online]. Available: http://www.ietf.org/mail-archive/web/ietf-announce/current/msg03407.html.

[130] A. Gupta, H. Sadawarti, and A. Verma, "Effect of Mobility Parameters on the Performance of AODV Routing Protocol," *ijnmt.com*, vol. 3, no. 1, pp. 14–20, 2012.

[131] M. Mahdi and T. Wan, "Performance Comparison of MANETs Routing Protocols for Dense and Sparse Topology," *ipcsit.com*, vol. 27, no. Icicn, pp. 37–41, 2012.

[132] C. Perkins and E. Royer, "Ad-hoc on-demand distance vector routing," ... *WMCSA'99. Second IEEE Workshop on*, 1999.

[133] I. GERASIMOV and R. SIMON, "Communication Support for Tightly-coupled Distributed Mobile Applications," *International Journal of Simulation*, pp. 23–39, 2002.

[134] A. Fehnker, R. V. Glabbeek, and P. Höfner, "Automated analysis of AODV using UPPAAL," ... *and Analysis of Systems*, 2012.

[135] M. K. Marina and S. R. Das, "Ad hoc on-demand multipath distance vector routing," *Wireless Communications and Mobile Computing*, vol. 6, no. 7, pp. 969–988, Nov. 2006.

[136] B. Kawish, B. Aslam, and S. Khan, "Reducing the Overhead Cost in Fixed & Low Mobility AODV Based MANET's," *Proceedings of the ...*, pp. 569–578, 1896.

[137] B. Kawish, B. Aslam, and S. Khan, "Reduction of Overheads with Dynamic Caching in Fixed AODV based MANETs," *World Academy of Science, ...*, pp. 840–845, 2006.

[138] M. Akhlaq, M. Jafri, M. Khan, and B. Aslam, "Addressing security concerns of data exchange in aodv protocol," *World Academy of Science, ...*, vol. 16, no. November, pp. 29–33, 2006.

[139] M. Murthy, S. Madhavi, G. Kumar, and Y. Kumar, "Impact of Node Convergence and Divergence in AODV and OLSR Routing Protocols," *ijcst.org*, vol. 3, no. 4, 2012.

[140] A. Thakare and M. Joshi, "Performance Analysis of AODV & DSR Routing Protocol in Mobile Ad hoc Networks," *International Journal of Computer …*, pp. 211–218, 2010.

[141] A. Solheid, "AODV enhanced by Smart Antennas," *fundp.ac.be*, 2005.

[142] S. D. Ubarhande, "Performance Evolution of AODV and DSR Routing Protocols in MANET Using NS2," vol. 3, no. 5, pp. 2–6, 2012.

[143] M. Oh, "An adaptive routing algorithm for wireless mesh networks," *… Communication Technology, 2008. ICACT 2008. 10th …*, no. x, pp. 2087–2091, Feb. 2008.

[144] C. Perkins, "Ad hoc On-Demand Distance Vector (AODV) Routing." [Online]. Available: http://www.ietf.org/rfc/rfc3561.txt.

[145] K. Chin and D. Lowe, "ROAR: a multi-rate opportunistic AODV routing protocol for wireless ad-hoc networks," *Ad-Hoc, Mobile, and Wireless Networks*, pp. 87–100, 2006.

[146] M. B. Yassein, M. O. Khaoua, L. M. Mackenzie, and S. Papanastasiou, "Performance Analysis of Adjusted Probabilistic Broadcasting in Mobile Ad Hoc Networks Performance Analysis of Adjusted Probabilistic Broadcasting in Mobile Ad Hoc Networks," no. 44.

[147] "Performance Evaluation of MANET Routing Protocols Using Scenario Based Mobility Models." [Online]. Available: http://books.google.co.uk/books?id=nH3F1UtwPCYC&pg=PA420&lpg=PA420&dq=The+destination+s

ends+the+RREP+as+a+unicast+message&source=bl&ots=jQKzR3JUy8&sig=2QONoqjDPjXDTk9DS3it5mwBeZw&hl=en&sa=X&ei=sK9bULu5MvLM0AW46IGIDQ&ved=0CDsQ6AEwBA#v=onepage&q=The destination sends the RREP as a unicast message&f=false.

[148] A. Amuthan, N. Sreenath, and S. Parthiban, “Proactive Secure Measures on Consolidated Query Multicast Protocol for Mobile Ad-hoc Networks,” *Int. J. Emerg. Sci*, vol. 2, no. September, pp. 478–498, 2012.

[149] A. H. Al-Bayatti, H. Zedan, and A. Cau, “Security Solution for Mobile Ad Hoc Network of Networks (MANoN),” *2009 Fifth International Conference on Networking and Services*, pp. 255–262, 2009.

[150] Z. Danailov, “Attacks on Mobile Ad hoc Netwoks.”

[151] J. Yu and H. Lee, “Traffic flooding attack detection with SNMP MIB using SVM,” 2008.

[152] B. Wu, J. Chen, J. Wu, and M. Cardei, “A Survey of Attacks and Countermeasures in Mobile Ad Hoc Networks.” [Online]. Available: http://www.citeulike.org/user/csands/article/9156912

[153] M. Sharma, N. Rizvi, N. Sharma, A. Malhan, and S. Sharma, “Performance Evaluation of MANET Routing Protocols under CBR and FTP traffic classes,” *International Journal*, vol. 2, no. 3, pp. 392–400, 2011.

[154] F. Schmidt, M. Alizai, I. AktaÅŸ, and K. Wehrle, “Refector: heuristic header error recovery for error-tolerant transmissions,” *Proceedings of the Seventh …*, 2011.

[155] S. Ali and A. Ali, "Performance Analysis of AODV, DSR and OLSR in MANET," pp. 1–59, 2010.

[156] Y. Zhang and L. Cheng, "Improving Reliability of Packet Delivery in MANETs by a Holistic Routing Approach," *Department of Electrical and Computer Engineering ...*, pp. 1–19, 2006.

[157] M. Lee and S. Lee, "A link stability model and stable routing for mobile ad-hoc networks," *Embedded and Ubiquitous Computing*, pp. 904–913, 2006.

[158] W. Wei and A. Zakhor, "Interference aware multipath selection for video streaming in wireless ad hoc networks," *Circuits and Systems for Video Technology, ...*, pp. 1–15, 2009.

[159] S. Sarkar, B. Kisku, S. Misra, and M. S. Obaidat, "Chinese Remainder Theorem-Based RSA-Threshold Cryptography in MANET Using Verifiable Secret Sharing Scheme," *IEEE International Conference on Wireless and Mobile Computing, Networking and Communications*, pp. 258–262, Oct. 2009.

[160] N. Bilandi and H. Verma, "Comparative Analysis of Reactive, Proactive and Hybrid Routing Protocols in MANET," *ijecse.org*, vol. 1, no. 3, pp. 1660–1667, 2012.

[161] R. Durán, I. D. Miguel, and N. Merayo, "Minimization of end-to-end delay in reconfigurable WDM Networks by means of genetic algorithms," *Submitted to Proc. ...*, no. May, pp. 22–24, 2006.

[162] D. Chaudhary, P. Panwar, and D. Waghmare, "Comparison and Performance Evaluation of Wireless Sensor Network with different Routing

Protocols," *International Conference on ...*, vol. 6, pp. 278–282, 2011.

[163] A. El Gamal, J. Mammen, and B. Prabhakar, "Throughput-Delay Trade-off in Wireless Networks," *Ieee Infocom 2004*, vol. 1, pp. 464–475, 2004.

[164] M. Mehla and R. Dahiya, "Performance Enhancement in Mobile Computing Using Replicated Cache Agent," *ijceronline.com*, vol. 2, no. 3, pp. 860–867, 2012.

[165] J. Mammen and D. Shah, "Throughput and delay in random wireless networks: 1-D mobility is just as good as 2-D," *International Symposium onInformation Theory, 2004. ISIT 2004. Proceedings.*, pp. 437–437, 2004.

[166] I. F. Akyildiz, W. Su, Y. Sankarasubramaniam, and E. Cayirci, "Wireless sensor networks: a survey," *Computer Networks*, Mar-2002. [Online]. Available: http://linkinghub.elsevier.com/retrieve/pii/S1389128601003024. [Accessed: 03-Sep-2012].

[167] J.-P. Yang, C.-K. Kung, F.-T. Liu, Y.-J. Chen, C.-Y. Chang, and R.-C. Hwang, "A Study of End-to-end Delay in MANET," *2010 First International Conference on Pervasive Computing, Signal Processing and Applications*, pp. 456–459, Sep. 2010.

[168] S. Cho and J. P. Hayes, "Impact of mobility on connection stability in ad hoc networks," *IEEE Wireless Communications and Networking Conference, 2005*, vol. 3, pp. 1650–1656, 2005.

[169] A. Al-Maashri and M. Ould-Khaoua, "Performance analysis of MANET routing protocols in the presence

of self-similar traffic," *Local Computer Networks, ...*, no. November, pp. 801–807, 2006.

[170] P. Naidu and M. Chawla, "Extended Ad Hoc on Demand Distance Vector Local Repair Trial for MANET," *International Journal of Wireless & Mobile Networks (...*, vol. 4, no. 2, pp. 235–250, 2012.

[171] T. El-Nabi and A. Ahmed, "Modeling and simulation of a routing protocol for Ad hoc networks combining queuing network analysis and Ant colony algorithms," *Institut für Informatik und Wirtschaftsinformatik Essen*, no. April, 2005.

[172] A. Radwan, T. Mahmoud, and E. Houssein, "Performance Measurement of Some Mobile Ad Hoc Network Routing Protocols," *International Journal of ...*, vol. 8, no. 1, pp. 107–112, 2011.

[173] M. Solimana and S. El-kaderb, "New adaptive routing protocol for MANET," *Ubiquitous Computing ...*, vol. 2, no. 3, pp. 16–23, 2007.

[174] D. Zamboni and C. Kruegel, "Recent Advances in Intrusion Detection," 2006. [Online]. Available: http://books.google.co.uk/books?id=9iv9UpdyjhUC&pg=PA266&lpg=PA266&dq=node+speed+is+increased+more+message+overhead+to+confront+these+changes.&source=bl&ots=RWGPVAtRU4&sig=Kj8v6FDOfVmSY5wztRAsOw7S-_8&hl=en&sa=X&ei=h1deULzBMfC20QWw14HoDA&ved=0CDAQ6AEwAA#v=onepage&q=node speed is increased more message overhead to confront these changes.&f=false.

[175] G. Kalpana and M. Punithavalli, "Reliable Broadcasting Using Efficient Forward Node Selection for Mobile Ad-Hoc Networks," *The International Arab Journal of Information ...*, vol. 9, no. 4, pp. 299–305, 2012.

[176] S. Bettuzzi, "Chapter 1: Introduction.," *Advances in cancer research*, vol. 104, pp. 1–8, Jan. 2009.

[177] S. Li, Q. Liu, H. Chen, and M. Tan, "A New Method to Resist Flooding Attacks in Ad Hoc Networks," *2006 International Conference on Wireless Communications, Networking and Mobile Computing*, pp. 1–4, Sep. 2006.

[178] J. W. Creswell, "Qualitative, Quantitative and Mixed Methods Approaches," 2002.

[179] E. Weingartner, "A performance comparison of recent network simulators," *... , 2009. ICC'09. IEEE ...*, 2009.

[180] "NS-3: Network Simulator 3." [Online]. Available: www.nsnam.org.

[181] J. Gustavo, "NS-3: Network Simulator 3." [Online]. Available: http://www.nsnam.org/tutorials/NS-3-LABMEETING-1.pdf.

[182] M. and S. Floyd, "The LBNL network simulator." [Online]. Available: http://www.isi.edu/nsnam.

[183] M. L. T. R. Henderson, "Network simulations with the ns-3 simulator."

Appendix

Instructions to Run the Code

1. Overwrite the AODV folder in the /ns-allinone-3.14.1/ns-3.14.1/src/aodv (depends where the ns-3 is installed).

2. Compile and run by ./waf --run "examples/aodv", you must be in main aodv folder.

3. To change any parameters open the aodv.cc file in aodv/examples folder and change the global variables.

Such as:

enable_flooding

flooding_intervel

flooding_broadcast

msg_broadcast

enable_malicious to enable Black Hole attack, disable flooding_enable when simulating Black Hole attack

Choose different options to run the system with all three scenarios given in the book and just make enable_flooding= false to run without any flooding.

4. To implement/de-implement prevention comment/uncomment the two if statements as shown below route input function in the aodv/model/aodv-routing-protocl.cc

// Uncoment the next if statement to apply FAP for flooding broadcast

```
if(udpHeader.GetDestinationPort()==100)
```

```
{
std::cout<<"Destination IP "<<dst<<" and Port is "<<udpHeader.GetDestinationPort()<<" Source IP is "<<origin<<"\n";
RecvError (packet, "255.255.255.255" );
NS_LOG_DEBUG ("Drop packet " << p->GetUid () << " because no route to forward it.");
return false;
}

//Uncomment the if statement to implement FAP for unicast Flooding attack
If (udpHeader.GetDestinationPort()==100)
{
std::cout<<"Unicast Destination IP "<<dst<<" and Port is "<<udpHeader.GetDestinationPort()<<" interface is "<<iif<<"\n";
RecvError (pack, "255.255.255.255");
return false;
}
```

United Scholars Publications

Email: info@unitedscholars.net
www.unitedscholars.net

Made in United States
Orlando, FL
04 November 2026

90201571R00107